BACKGROUND MATH FOR A
COMPUTER WORLD

D1533324

BACKGROUND MATH FOR A COMPUTER WORLD

RUTH ASHLEY

University of Michigan
Ann Arbor, Michigan

in consultation with

Nancy B. Stern

Suffolk Community College

John Wiley & Sons, Inc.

New York · London · Sydney · Toronto

Library of Congress Cataloging in Publication Data

Ashley, Ruth.
    Background math for a computer world.

    (Wiley self-teaching guides)
    1. Mathematics--1961-   2. Mathematics--
Programmed instruction.   3. Electronic data process-
ing--Mathematics.   I. Title
QA39.2.A75   510'.7'7   72-8948
ISBN 0-471-03506-8

Printed in the United States of America

73   74   10 9 8 7 6 5

To the Reader

This book is not designed for mathematicians. It is prepared for the tens of thousands of people who find that their lives are being increasingly affected by computers. It is for the students with no college math and very limited high school mathematics who discover that they, too, are expected to be able to use computers—in business, in psychology, in education, in the social sciences. For these students this book will fill a critical need—the need to know enough mathematics to survive and be effective in the computer world.

This book presents basic facts and techniques from many different topic areas. The topics selected were agreed upon by experts in the computer field as representing the areas where knowledge is especially important to a beginner in computer work. After you complete this book you will not be an expert in any area, but you will be able to function, to understand what is going on, and to identify where you need more information. You will know that you can learn math, even if you have avoided it since junior high school. Several times on each page you will know if you are learning the material. At the end of each chapter you will prove to yourself that you have learned, and eventually you will verify this proof with the Final Test. The real test, of course, is whether your new knowledge is applied as your computer exposure grows. This part is up to you.

Keep in mind that this book is a background for computer work. It is not intended to be everything you will ever need to know about mathematics. But it will provide you with a good beginning.

The concepts included in the book do not reflect any particular programming language; they will be equally useful if you learn FORTRAN, COBOL, PL/I, BASIC, or any machine language for programming computers. The knowledge and skills you acquire here will be equally applicable in a gigantic computer installation or in a small operation. With the appropriate background in the fundamentals of mathematics, your encounters with computers will have more meaning, be more effective, and be more satisfying to you. You will be better able to keep abreast of modern developments in the computer world and to see how these developments can be applied to your own profession.

How to Use This Book

Each chapter in this book consists of many small segments called frames. Each frame includes a question to answer or a problem to solve and is followed by the correct answer or solution. When you work through this self-teaching guide, use a piece of paper or a large index card to cover the answers. You can slide the card or paper down to the dotted line. Next read the frame and write down your answer. Then slide the card or paper down again to check your response against the given answer. Most of the time you will be right. When you do find that you have made an error, find out why before going on. Look back over the preceding frames and make sure you understand the problem.

Each chapter will take from one to three hours to complete. Whenever possible, try to complete a chapter in one or two study sessions. Stop only at the end of a chapter or a section within the chapter. The material will be more difficult for you to learn if your work is frequently interrupted.

The computations in this book are kept to a minimum, but they cannot be avoided altogether. They can all be done by hand. If you have access to an adding machine or calculator, feel free to use it. Such devices should not hamper your learning of the concepts presented in this book.

Each chapter is followed by a Self-Test. Take the test after you have completed the chapter, then check your answers. If you make a mistake, check the solution as given in the answer key and then refer to the frames associated with that item. If you should miss more than half the questions in any Self-Test, you may find it helpful to repeat that chapter.

You may also use the Self-Test to discover if you already know the material covered in a chapter. Students with some math background may want to take the test first and study the chapter only if they did poorly on the test. In this way, they can skip ahead and avoid repeating material they already know.

PREREQUISITES

The first nine chapters of this self-teaching guide can be mastered with very minimal mathematics background. You should be able to:

- add and subtract decimal numbers and fractions, both positive and negative;

- multiply and divide with reasonable accuracy;

- solve very basic algebraic equalities, such as $x + 4 = 7$ and $6y = 20$.

The last three chapters require a tiny bit more. A moderate knowledge of high school algebra would be quite adequate. If you feel you need extra review in this area, acquire any first-year algebra book and refresh your knowledge.

Contents

BACKGROUND MATH FOR A COMPUTER WORLD

CHAPTER ONE
The Binary Number System

People in the modern world usually count and calculate using the decimal
—or base-ten—number system. The very early digital computers also
used decimal numbers, but this was soon found to be wasteful of both
equipment and time. Today the decimal system is not used directly in
any digital computers. Instead the base-two system is used almost ex-
clusively. Before looking at the structure of the base-two number system, it may be useful to review the decimal system, for the basic structure remains the same in both systems.

When you complete this chapter, you will be able to:

- convert decimal numbers to the binary number system;

- convert binary numbers to their decimal equivalents;

- add any two binary numbers;

- subtract any two binary numbers using the complement
  method;

- multiply any two binary numbers.

Prerequisite mathematics for the first two chapters are minimal.
You need to be able to add, subtract, and multiply decimal numbers,
both with and without decimal points.

REVIEW OF THE DECIMAL NUMBER SYSTEM

1. The decimal system uses ten different digits. These are 0, 1, 2, 3, 4, 5, 6, 7, 8, and 9. After these ten digits are each used once, you would write the next number by (check one):

    ___(a) using a new digit
    ___(b) writing a two-digit number
    ___(c) placing a "1" in the left position and starting over with the first of the original ten digits in the right position

    _ _ _ _ _ _ _ _ _ _ _ _ _ _

    (b); (c) (10 is a two-digit number.)

2. Each position in a decimal number has a value attached to it.

1000	100	10	1	.1

    9783.6

    9   7   8   3   .6

    In the example above, the value associated with the leftmost position is:

    ___(a) 1000
    ___(b) .1
    ___(c) 10

    _ _ _ _ _ _ _ _ _ _ _ _ _ _

    (a)

3. The position of each digit gives the value associated with it. Each position is valued at ten times the value of the right-hand position.

    If the rightmost position is 1, the next position on the left is _____.

    _ _ _ _ _ _ _ _ _ _ _ _ _ _

    10

4. The decimal value 237 can be written in an expanded form to show the importance of position.

    $$237 = (2 \times 100) + (3 \times 10) + (7 \times 1)$$
    $$\text{or } 237 = 200 + 30 + 7$$

    Write 1798 in expanded form. _____

    _ _ _ _ _ _ _ _ _ _ _ _ _

    $1798 = (1 \times 1000) + (7 \times 100) + (9 \times 10) + (8 \times 1)$

5. The expanded form of a decimal can be written more easily by using the powers of ten. Any number raised to a power is called a base.

   Thus, another name for the decimal system is the base-_____ system.

   - - - - - - - - - - - - - - -

   ten

6. The values of some of the powers of ten are probably known to you. The value of 10^2, for example, is 10 x 10, or 100. The value of 10^3 is 1000. Which of the choices below is the value of 10^4? (More than one can be right.)

   ___(a) 10 x 4
   ___(b) 10 x 10 x 10 x 10
   ___(c) 4 x 4 x 4 x 4 x 4 x 4 x 4 x 4 x 4 x 4
   ___(d) 10,000

   - - - - - - - - - - - - - - -

   (b); (d)

7. The value of 10^2 is 10 x 10. The value of 10^1 is simply 10. Any number raised to the first power has the same value as that base. 17^1, for example, has a base of 17. It is raised to the first power, so the value of 17^1 is 17. Write the value of each of these numbers:

   (a) 2^1 _____

   (b) 5^1 _____

   (c) 8^1 _____

   - - - - - - - - - - - - - - -

   (a) 2; (b) 5; (c) 8

8. A base raised to the zero power is defined as equal to 1. 2^0, for example, equals 1. The value of 5×10^0 is _____.

   - - - - - - - - - - - - - - -

   5 (or 5 x 1)

9. Write the values of each of the following:

   (a) 10^1 = _____   (b) 10^0 = _____

   (c) 2×10^0 = _____   (d) 4×10^1 = _____

- - - - - - - - - - - - - - -

(a) 10; (b) 1; (c) 2 (2 x 1 = 2); (d) 40 (4 x 10 = 40)

10. What is the value of $(6 \times 10^1) + (4 \times 10^0)$? _____

- - - - - - - - - - - - - - -

64 [(6 x 10) + (4 x 1) = 60 + 4]

11. What is the value of $(7 \times 10^2) + (1 \times 10^1)$? _____

- - - - - - - - - - - - - - -

710 [(7 x 100) + (1 x 10) = 710]

12. If you were expanding a number such as 710 you might write
$(7 \times 10^2) + (1 \times 10^1) + (0 \times 10^0)$. This is also correct, of course,
since zero times another number always equals zero. It is often
useful to think of the zero in a number as being the base raised to
the appropriate power times zero.
   Write out the expanded form of 9003, using four different powers

of 10. _____

- - - - - - - - - - - - - - -

$9003 = (9 \times 10^3) + (0 \times 10^2) + (0 \times 10^1) + (3 \times 10^0)$

13. Any decimal number, no matter how large or small, can be expressed as the sum of powers of ten. For example,

$$3,493,712 = (3 \times 10^6) + (4 \times 10^5) + (9 \times 10^4) + (3 \times 10^3) +$$
$$(7 \times 10^2) + (1 \times 10^1) + (2 \times 10^0)$$

The base, 10, is raised to one higher power for each position move
to the left.
   In the number 4,567,890,123 what power of ten is multiplied by

the "5"? _____

- - - - - - - - - - - - - - -

10^8

14. For the numbers listed below, write the power of ten that is represented by the position occupied by the digit "4."

(a) 34,271 _____

(b) 47,000,000 _____

(c) 4001 _____

- - - - - - - - - - - - - - -
(a) 10^3; (b) 10^7; (c) 10^3

15. Decimal fractions, such as .1, .4, and 1.987, are also represented as powers of ten. The power of ten is decreased by one for each position shift to the right. Thus, the position to the immediate right of 10^1 is 10^0. The position to the immediate right of 10^0 is 10^{-1}.
    What is the value of the position to the immediate right of 10^{-1}?

_____
- - - - - - - - - - - - - -
10^{-2}

16. In a decimal fraction, the decimal point comes immediately to the right of the 10^0 position. (The decimal point isn't always included with whole numbers, but it is assumed to be there.) The decimal 23.7 is equal to $(2 \times 10^1) + (3 \times 10^0) + (7 \times 10^{-1})$, which indicates that the value of 10^{-1} is one-tenth.
    The value of a number with a "minus" exponent (power) is one divided by the base with the "plus" exponent, or the reciprocal of the "plus" exponent. For example, 10^{-1} is equal to $1/10^1$, or $1/10$, or in the usual form, .1.
    What is the value of 10^{-2}? _____
- - - - - - - - - - - - - -
$10^{-2} = 1/10^2 = 1/100 = .01$ ($1/10^2$ is the reciprocal of 10^{-2}.)

17. Write the fractions that represent the following values. Use the form given for (a).
    (a) $10^{-2} = $ ___1/100___   (b) $10^{-3} = $ _____
    (c) $10^{-7} = $ _____   (d) $10^{-5} = $ _____
- - - - - - - - - - - - - -
(b) 1/1000; (c) 1/10,000,000; (d) 1/100,000

18. Write each of the following as a power of ten.
    (a) $1/100 = $ _____   (b) $1/10 = $ _____
    (c) $1000 = $ _____   (d) $1/10,000 = $ _____
- - - - - - - - - - - - -
(a) 10^{-2}; (b) 10^{-1}; (c) 10^3 (or 10^{+3}); (d) 10^{-4}

19. Any decimal can be written in expanded form using the powers of ten. The position immediately to the left of the decimal point is always 10^0. The position immediately to the right of the decimal point is always 10^{-1}. The value of each position is always one larger or one smaller than the adjacent position.

Now write the expanded form of 5,719.204.

- - - - - - - - - - - - - - -

$(5 \times 10^3) + (7 \times 10^2) + (1 \times 10^1) + (9 \times 10^0) + (2 \times 10^{-1}) + (0 \times 10^{-2}) + (4 \times 10^{-3})$

The (0×10^{-2}) may be omitted, since it is equal to zero.

BINARY NUMBERS

You have seen the basic structure of the base-ten number system. You know that:

- a minus exponent means a reciprocal ($10^{-2} = 1/10^2$);

- any number raised to the zero power is equal to one ($10^0 = 1$);

- any number raised to the first power is equal to that number ($10^1 = 10$).

These facts hold true for other number systems too.

Next we are going to study the binary number system, which has the base two. Most computer systems in use today depend on the binary system for their speed and efficiency in calculations. Although the binary system gets quite awkward in paper-and-pencil work, a thorough understanding of how it works is vital for the efficient use of computers.

Computers are composed of millions of electrical circuits. Each circuit can be either open or closed. (These conditions are sometimes called off and on.) So each circuit has two basic states: conducting electricity (closed or on) and not conducting electricity (open or off). These two states can easily represent the binary digits. Binary is a base-two number system, which uses just two digits, 1 (on) and 0 (off), to represent all values.

20. 101_2 is a binary number; it is read "one-zero-one, base two." Since it is a base-two number, it can be expanded using <u>powers of two</u>. The rightmost position here represents 2^0, the next left position represents 2^1, and the next position on the left represents 2^2. The expanded form would be written:

$$101_2 = (1 \times 2^2) + (0 \times 2^1) + (1 \times 2^0)$$

Complete the expansion below to find the decimal equivalent of 101_2.

(a) $2^2 =$ _____ , so $1 \times 2^2 =$ _____

(b) $2^1 =$ _____ , so $0 \times 2^1 =$ _____

(c) $2^0 =$ _____ , so $1 \times 2^0 =$ _____

(d) Add the totals in the
final column to find
the decimal equivalent: _____

- - - - - - - - - - - - - -

(a) 4, 4; (b) 2, 0; (c) 1, 1; (d) 5

21. Any binary number is a string of zeros and ones; any binary number can be converted to its decimal equivalent as illustrated in frame 20. But first we need to know the decimal values of the powers of two. In the last frame, you knew the value of 2^0 and 2^1. You also knew that 2^2 is equal to 2×2, or 4. The value of 2^8 is $2 \times 2 \times 2 \times 2 \times 2 \times 2 \times 2 \times 2$, or the product of eight twos, which is 256.
Find the decimal value of each of these powers of two.

(a) $2^3 =$ _____

(b) $2^4 =$ _____

(c) $2^6 =$ _____

(d) $2^7 =$ _____

- - - - - - - - - - - - - -

(a) 8; (b) 16; (c) 64; (d) 128

22. If you are not familiar with the powers of two, you will find it helpful to make a small chart, giving these equivalents. You will learn them in time, but a chart will make it easier than recalculating the values each time you need them. Just copy this one.

2^0	2^1	2^2	2^3	2^4	2^5	2^6	2^7	2^8	2^9
1	2	4	8	16	32	64	128	256	512

In expanding binary numbers to find their decimal equivalents, it is useful to use a layout like the one shown below. Fill in the missing values, then find the sum, which will be the decimal equivalent of 1100110_2.

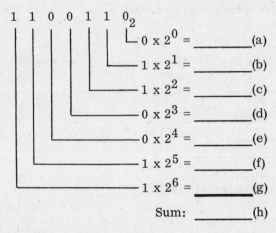

$$0 \times 2^0 = \underline{\hspace{1cm}} \text{(a)}$$
$$1 \times 2^1 = \underline{\hspace{1cm}} \text{(b)}$$
$$1 \times 2^2 = \underline{\hspace{1cm}} \text{(c)}$$
$$0 \times 2^3 = \underline{\hspace{1cm}} \text{(d)}$$
$$0 \times 2^4 = \underline{\hspace{1cm}} \text{(e)}$$
$$1 \times 2^5 = \underline{\hspace{1cm}} \text{(f)}$$
$$1 \times 2^6 = \underline{\hspace{1cm}} \text{(g)}$$
$$\text{Sum:} \underline{\hspace{1cm}} \text{(h)}$$

- - - - - - - - - - - -

(a) 0; (b) 2; (c) 4; (d) 0; (e) 0; (f) 32; (g) 64; (h) 102

23. Expand the binary number 11011_2, using the powers of two.

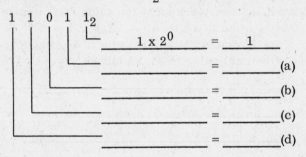

$$\underline{\hspace{1.5cm}1 \times 2^0\hspace{1.5cm}} = \underline{\hspace{0.5cm}1\hspace{0.5cm}}$$
$$\underline{\hspace{3cm}} = \underline{\hspace{1cm}} \text{(a)}$$
$$\underline{\hspace{3cm}} = \underline{\hspace{1cm}} \text{(b)}$$
$$\underline{\hspace{3cm}} = \underline{\hspace{1cm}} \text{(c)}$$
$$\underline{\hspace{3cm}} = \underline{\hspace{1cm}} \text{(d)}$$

- - - - - - - - - - - -

(a) $1 \times 2^1 = 2$; (b) $0 \times 2^2 = 0$; (c) $1 \times 2^3 = 8$; (d) $1 \times 2^4 = 16$

24. Add the expanded values from frame 23 to give the decimal equivalent of 11011_2. \underline{\hspace{6cm}}

- - - - - - - - - - - -

27 $(1 + 2 + 0 + 8 + 16 = 27)$

25. Complete each line of the following layout in order to find the decimal equivalent of 1001_2.

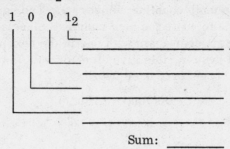

Sum: _____

- - - - - - - - - - - - - - -

$1001_2 = (1 \times 2^0) + (0 \times 2^1) + (0 \times 2^2) + (1 \times 2^3) = 9_{10}$

26. Find the decimal equivalent of 11111000_2.

$1\ \ 1\ \ 1\ \ 1\ \ 1\ \ 0\ \ 0\ \ 0_2$

- - - - - - - - - - - - - - -

248 $(0 + 0 + 0 + 8 + 16 + 32 + 64 + 128)$

BINARY COUNTING

You can now convert any whole binary number into its decimal equivalent. Next you are going to learn to count in binary. You already know the first two counting numbers because they are the same as in the decimal system: 0 and 1. You will probably memorize the binary equivalents of the first ten decimal counting numbers as you work through the following frames.

27. After the first two binary counting numbers, 0 and 1, all of the binary digits are used up. In the decimal system, we used up all the digits after the tenth counting number, 9. The same method is used in both systems to write the next number: place a zero in the "ones" position and start over again with one in the next position to the left. In the decimal system, this gives ten, or 10. In binary it gives 10_2, which is read "one-zero, base two."

Decimal	Binary
0	0
1	1
2	10

Write out the expanded form of 10_2 and use powers of two to verify that it is equivalent to the decimal 2.

$10_2 = $ _____

- - - - - - - - - - - - - -

$10_2 = (0 \times 2^0) + (1 \times 2^1) = 0 + 2 = 2$

28. When counting, we mentally add 1 to each number to get the next number. In the decimal system, $1 + 1 = 2$, the number that follows 1. In binary, $1 + 1 = 10_2$, so 10 is the number that follows 1 in the binary system.

Look back at the first three decimal-binary equivalents in frame 27. What is the value of $0 + 1$ in the binary system? _____

- - - - - - - - - - - - - -

$0 + 1 = 1$ (The number that follows 0 is 1.)

29. Adding is done by position in binary, just as it is in decimal addition. In order to get the binary equivalent of the decimal 3, we can add 1 to the binary equivalent of the decimal 2. You know that $0 + 1 = 1$ in binary. What is the value of $10_2 + 1_2$?

$$\begin{array}{r} 10_2 \\ + \ 1_2 \\ \hline \end{array}$$

- - - - - - - - - - - - - -

11_2 (This is read "one-one, base two.")

30. Addition in binary is quite easy to learn since there are only four possible combinations of digits to be added. These can be represented in a table.

Binary Addition Table

+	0	1
0	0 + 0 0	0 + 1 1
1	1 + 0 1	1 + 1 10

Use the binary addition table above to give the following sums.

(a) $1 + 0 =$ _____   (b) $0 + 1 =$ _____

(c) $1 + 1 =$ _____   (d) $0 + 0 =$ _____

- - - - - - - - - - - - - - - -

(a) 1; (b) 1; (c) 10; (d) 0

31. In binary addition, as in decimal addition, it is frequently necessary to "carry" a 1. You already know that the decimal 3 is equivalent to 11 in binary. The decimal 4 is then equivalent to the binary sum of 11 + 1. To keep the positions straight, we shall stack these numbers.

$$\begin{array}{r} 11_2 \\ + \ 1_2 \\ \hline \end{array}$$ To find the sum, add the right-hand column first. According to the binary addition table, 1 + 1 = 10. Write down 0 below the right-hand column, and a small 1 above the left-hand column.

- - - - - - - - - - - - - -

$$\begin{array}{r} {}^{1}11_2 \\ + \ 1_2 \\ \hline 0 \end{array}$$

32. $$\begin{array}{r} {}^{1}11_2 \\ + \ 1_2 \\ \hline 0 \end{array}$$ Now add the remaining column. Find 1 + 1 on the binary addition table, and write it down, since this is the last column. Now you have found the binary equivalent of decimal 4.

- - - - - - - - - - - - - -

11_2
$+ \ 1_2$
———
100_2

33. Write the first five binary numbers, starting with the equivalent of the decimal 0. _____

- - - - - - - - - - - - - -

$0_2, \ 1_2, \ 10_2, \ 11_2, \ 100_2$

34. If the binary equivalent of decimal 4 is 100_2, then the binary equivalent of decimal 5 is $100_2 + 1_2$, which is _____.

- - - - - - - - - - - - - -

101_2

35. Find the binary equivalent of decimal 6 and decimal 7 by adding 1 to 101_2 two times.

Decimal	Binary
5	101_2
6	_____
7	_____

- - - - - - - - - - - - - -

$6 = 110_2; \ \ 7 = 111_2$

36. The addition required to find the binary equivalent of decimal 8 is a bit more complicated, but it can be done by using the binary addition table in frame 30 and carrying carefully. Try it.

111_2
$+ \ \ \ 1_2$

- - - - - - - - - - - - - -

111_2
$+ \ \ \ 1_2$
———
1000_2

37. Fill in the missing binary values in the chart below.   You will probably have to recalculate a few of them.

Decimal	Binary
0	_____
1	1
2	_____
3	_____
4	_____
5	_____
6	110
7	_____
8	_____
9	1001
10	_____

- - - - - - - - - - - - - - -

Decimal	Binary
0	0
1	1
2	10
3	11
4	100
5	101
6	110
7	111
8	1000
9	1001
10	1010

You now know how to count in binary and how to add one to any number. You can also convert numbers from binary to their decimal equivalents. The next step is to learn to convert decimals into their binary equivalents without having to count. You do this by using the familiar decimal arithmetic, subtracting the values of the different powers of two. Do you still have the chart you copied from frame 22? The values of the powers of two are used in decimal to binary conversion. Below is an example of the decimal number 24 converted to binary.

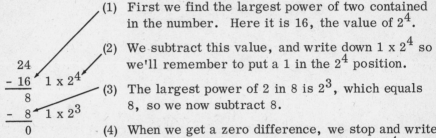

(1) First we find the largest power of two contained in the number. Here it is 16, the value of 2^4.

(2) We subtract this value, and write down 1×2^4 so we'll remember to put a 1 in the 2^4 position.

(3) The largest power of 2 in 8 is 2^3, which equals 8, so we now subtract 8.

(4) When we get a zero difference, we stop and write the binary result. Here we have a 1 for 2^4, a 1 for 2^3, and we will put 0 in all remaining positions. Thus the binary equivalent of decimal 24 is 11000.

38. The rules to follow in converting numbers from decimal to binary are:

- Subtract the value of the largest power of two contained in the decimal number.

- Repeat subtracting the value of the largest power of two from each result until you get a zero difference.

- Write a one for each position that you made a subtraction, and a zero for other positions.

Use this method to find the binary equivalent of decimal 15. (Decimal 15 can also be written 15_{10}.)

– – – – – – – – – – – – – –

$15_{10} = 1111_2$

$$
\begin{array}{rl}
15 & \\
-\ 8 & 1 \times 2^3 \\
\hline
7 & \\
-\ 4 & 1 \times 2^2 \\
\hline
3 & \\
-\ 2 & 1 \times 2^1 \\
\hline
1 & \\
-\ 1 & 1 \times 2^0 \\
\hline
0 &
\end{array}
$$

39. Find the binary equivalent of decimal 21.

– – – – – – – – – – – – –

$21_{10} = 10101_2$ (You subtracted 16, 4, and 1.)

40. Convert decimal 49 to a binary number.

– – – – – – – – – – – – –

$49_{10} = 110001_2$ (You subtracted 32, 16, and 1.)

41. Now convert the following decimals to binary. Remember the rules:

     • Subtract largest powers of two.
     • Write one for the positions you subtract.
     • Write zero for the other positions.

(a) 32_{10} (b) 100_{10}

(c) 892_{10}

– – – – – – – – – – – – – –

(a) 100000_2 (You subtracted only 32.)
(b) 1100100_2 (You subtracted 64, 32, and 4.)
(c) 1101111100_2 (You subtracted 512, or 2^9, 256, 64, 32, 16, 8, and 4.)

BINARY ADDITION

You can now convert decimal numbers to binary numbers, and binary numbers to decimal numbers. You can count in binary, and add one to any binary number. Now you are going to learn to add larger numbers. You will add only two numbers at a time, although occasionally a carried one will be included. Computers usually use an addition table like the one we used first in frame 30. Feel free to use a similar table if you wish. You will know the combinations very soon, however. When it is very clear that we are dealing with binary numbers, we shall omit the subscript 2.

42. You know that in decimal numbers 003 = 3. These "leading" zeros do not affect the value, but may be used for convenience. Leading zeros are usually used in binary addition. These do not, of course, change the value of the numbers; they do simplify addition. Referring to the binary addition table in frame 30 when necessary, find the following sums.

    (a)   001   (b)   011   (c)   100
        + 110   + 010   + 101

— — — — — — — — — — — — —

(a) 111; (b) 101; (c) 1001

43. You carried one in binary counting by writing a zero in the column and a one above the next left column. This procedure occasionally results in your having to add three ones. In binary, $1 + 1 + 1 =$ ____ .

— — — — — — — — — — — — —

11_2

44. Add the binary numbers below. Remember to use the binary addition table and write down the carried ones.

$$0111$$
$$+ 1011$$

— — — — — — — — — — — — —

  $^{I\,I\,I}$
0111
+ 1011
10010

45. Find these binary sums.

    (a)   1010   (b)   1110   (c)   11111
        + 0111   + 0110   + 01011

— — — — — — — — — — — — —

(a) 10001; (b) 10100; (c) 101010

46. You can now add together any two binary whole numbers. The method of adding binary numbers with "binary points" is exactly the same. The positions to the right of the binary point represent minus powers of two, or $\frac{1}{2}$, $\frac{1}{4}$, and so on.

Add the following binary numbers.

(a) 11.01
    + 01.10

(b) 10000.1100
    + 00000.1011

_ _ _ _ _ _ _ _ _ _ _ _ _ _ _

(a) 100.11; (b) 10001.0111

BINARY SUBTRACTION

Binary subtraction, like decimal subtraction, can be performed by using the binary addition table in reverse, but you need to use a borrowing technique. It becomes very involved and most people, like most computers, perform binary subtraction using a technique called complementing. The computer uses the complement of the smaller number, then adds the two together before shifting a one to a different position.

47. The binary complement of a number is written simply by writing a zero where there is a one, and a one where there is a zero. The complement of 101, for example, is 010. Write the complements of the following binary numbers.

(a) 110011 _____

(b) 111110 _____

(c) 01010 _____

_ _ _ _ _ _ _ _ _ _ _ _ _ _

(a) 001100; (b) 000001; (c) 10101

48. The complement in the binary system is actually the difference between each of the digits in a binary number and one. $1 - 0 = 1$, and $1 - 1 = 0$. In effect, you simply write down the opposite digit for each position. Try a few more complements.

(a) 000111 _____

(b) 111111 _____

(c) 1 _____

_ _ _ _ _ _ _ _ _ _ _ _ _ _

(a) 111000; (b) 000000; (c) 0

49. In binary subtraction you use the complement of the number you are
subtracting. Thus, if you had a problem like 1100 − 11, you would
add the complement of 11 to 1100.
     Rewrite the following problem, using the complement; then add
the two figures. Be sure to write in leading zeros to 11 before you
complement it.

$$
\begin{array}{r}
1100 \\
-\ \ \ 11 \\
\end{array}
$$

- - - - - - - - - - - - - -

$$
\begin{array}{r}
1100 \\
+\ 1100 \\
\hline
11000 \\
\end{array}
$$

50. The three steps in binary subtraction are:

           Step 1: complement
           Step 2: add
           Step 3: shift

By shift, we mean remove the leftmost one from your answer in
step 2 and add it to the rightmost column.
     What is the result when you apply step 3 to your answer to the

problem in frame 49? _____

- - - - - - - - - - - - - -

1001   (1)1000   (This shift is called an "end-around carry" in
          +   1   computer terminology.)
            ─────
             1001

51. The original subtraction problem in frame 49 looked like this:

$$
\begin{array}{r}
1100 \\
-\ 0011 \\
\end{array}
$$

We rewrote it with the complement of 0011 (step 1). Then we added
(step 2), and finally shifted the one (step 3).
     You can check the answer to see if it is correct, just as you
check decimal subtraction problems. Add the result obtained in
frame 50 to 0011.

(a) What is your answer? _____

(b) Does the problem check? _____

- - - - - - - - - - - - - -

(a) 1100; (b) yes (So the complement method does give the correct answer, and computers do it very efficiently.)

52. Now do a binary subtraction, step by step.

$$11001$$
$$- 01101$$

(a) Step 1: Rewrite the problem with the proper number complemented.

(b) Step 2: Add the two numbers together.

(c) Step 3: Shift the leftmost one and add it on to the right.

(d) Check your answer by adding.

- - - - - - - - - - - - -

(a)   11001   (b) 101011   (c) ⓵01011   (d)   1100
      + 10010   +   1   + 1101
                                            1100   11001

53. Remember the steps in binary subtraction.

   • Complement
   • Add
   • Shift
   • Check (This isn't really a step, but it is useful.)

Now try a few subtractions of binary numbers.

(a)   1111   (b)   1001   (c)   111001
      - 100   - 110   - 11010

- - - - - - - - - - - - - - -

(a)   1111
     + 1011   (complement)
     ⑪1010 (add)
  +   ↖1 (shift)
     1011 (answer)

(b)   1011
     + 1001   (complement)
     ①0010 (add)
  +   ↖1 (shift)
       11 (answer)

(c)   111001
    + 100101   (complement)
    ①0011110 (add)
  +   ↖1 (shift)
     11111 (answer)

54. Now try a subtraction with a "binary point." The method is identical to that used with whole numbers.

$$1110.101$$
$$-\ \ \ 11.011$$

- - - - - - - - - - - - - -

    1110.101
  + 1100.100 (complement)
  ①1011 011 (add)
  +   →1 (shift)
    1011.010 (answer)

BINARY MULTIPLICATION

55. Computers perform binary multiplication problems using a table, just as they do in addition. The binary multiplication table is very simple. Any binary number times zero equals zero. Any binary number times one equals the same number. A binary multiplication table is given below.

Binary Multiplication Table

x	0	1
0	0 x 0 0	0 x 1 0
1	1 x 0 0	1 x 1 1

Refer to the table and the information above to give the following simple products.

(a) 1 x 0 =   (b) 111 x 0 =

(c) 1 x 1 =   (d) 1 x 1011 =

- - - - - - - - - - - - - - -

(a) 0; (b) 0; (c) 1; (d) 1011

56. In binary multiplication, partial products are shifted over to keep the positions lined up, just as in decimal multiplication. Below is an example of binary multiplication.

$$
\begin{array}{r}
111 \\
\times\ 10 \\
\hline
000 \quad (1) \\
111 \quad (2) \\
\hline
1110 \quad (3)
\end{array}
$$

(1) Zero times any number equals zero, so we write zeros.
(2) One times any number equals that number, so we copy it. We shift the result over one column so the number begins under the one we are multiplying by.
(3) We add the two "partial products" to get the answer.
Use this method to find the following product.

$$
\begin{array}{r}
1101 \\
\times\ \ 10
\end{array}
$$

```
  1101
x   10
 0000
 1101
11010
```

57. As in decimal multiplication, it isn't necessary to write a string of zeros when multiplying by zero. You may write down just one zero, as shown here:

```
  111
x  10
 1110
```

Try these multiplication problems.

(a)   1101   (b)   11100
       x   10   x   100

(a) 11010; (b) 1110000

58. When multiplying two large binary numbers, you will have columns of many ones to add up. Since adding is much easier with only two digits, computers calculate "partial sums" and add them often as shown below.

```
      110011
    x  11101
      110011
     1100110    (One zero written for the zero product)
     11111111   (First partial sum)
    110011
    1010010111  (Second partial sum)
    110011
    10111000111 (Final answer)
```

Use the method illustrated above to find the following product.

```
     1100
   x  111
```

```
- - - - - - - - - - - -
    1100
  x  111
    1100
    1100
  100100
   1100
  1010100
```

59. Now do these binary multiplication problems.

    (a)   1101   (b)   1101   (c)   110011

        x   101   x 1111   x 110110

- - - - - - - - - - - - - -

    (a) 1000001; (b) 11000011; (c) 101011000010

60. Multiplication of binary numbers with "binary points" follows the same rules as those with whole numbers. As in decimal multiplication, the total number of places in the original two numbers appears in the result. Try this multiplication problem.

$$\begin{array}{r} 111.001 \\ \text{x} \quad 100.1 \\ \hline \end{array}$$

- - - - - - - - - - - - - -

100000.0001

Most computers perform binary division by using the method of repeated subtractions. This is very time-consuming on paper, so we will not study it. Since computers subtract by complementing, you have seen how all basic arithmetic is reduced to adding by a computer.

Computers take in decimal numbers and instructions to perform certain operations. They convert the decimals into binary numbers to store them, perform the operations, then reconvert the results to decimal numbers. It sounds reasonable and possible, but problems can arise in the process, especially when fractions are involved. For example, decimal 0.1 has no exact binary equivalent. The closest is 0.0001100110011. . ., repeating 0011 forever. An understanding of binary operations is helpful, though. If enough decimal places are used, and enough positions reserved for binary places, many problems become negligible.

You have learned to convert numbers from binary to decimal, and from decimal to binary. You can count in binary. You can add, subtract, and multiply binary numbers. Now take the Self-Test on the next page.

SELF-TEST

This Self-Test will help you evaluate whether or not you have mastered the chapter objectives and are ready to go on to the next chapter. Answer each question to the best of your ability. Correct answers are given at the end of the test.

1. Convert the following binary numbers to decimal equivalents.

   (a) 1101110_2

   (b) 111011_2

2. Convert the following decimal numbers to binary equivalents.

   (a) 270_{10}

   (b) 94_{10}

3. Solve the following binary problems.

    (a)   1110   (b)   111000   (c)   110011
           + 1010   -   111   x   1110

ANSWERS TO SELF-TEST

Compare your answers to the Self-Test with the correct answers given below. If all of your answers are correct, you are ready to go on to the next chapter. If you missed any questions, study the frames indicated in parentheses following the answer.

1. (a) 110

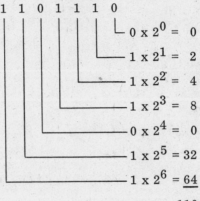

$$1 \; 1 \; 0 \; 1 \; 1 \; 1 \; 0$$

$0 \times 2^0 = 0$

$1 \times 2^1 = 2$

$1 \times 2^2 = 4$

$1 \times 2^3 = 8$

$0 \times 2^4 = 0$

$1 \times 2^5 = 32$

$1 \times 2^6 = \underline{64}$

110   (frames 22-26)

(b) 59

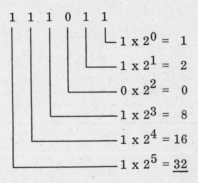

$$1 \; 1 \; 1 \; 0 \; 1 \; 1$$

$1 \times 2^0 = 1$

$1 \times 2^1 = 2$

$0 \times 2^2 = 0$

$1 \times 2^3 = 8$

$1 \times 2^4 = 16$

$1 \times 2^5 = \underline{32}$

59   (frames 22-26)

2. (a) 100001110

$$\begin{array}{r} 270 \\ -\,256 \\ \hline 14 \end{array} \quad 1 \times 2^8$$

$$\begin{array}{r} -\;8 \\ \hline 6 \end{array} \quad 1 \times 2^3$$

$$\begin{array}{r} -\,4 \\ \hline 2 \end{array} \quad 1 \times 2^2$$

$$\begin{array}{r} -\,2 \\ \hline 0 \end{array} \quad 1 \times 2^1$$

(frames 38-41)

(b) 1011110   94
                   $- 64$ 1×2^6
                    30
                   $- 16$ 1×2^4
                    14
                   $- 8$ 1×2^3
                     6
                   $- 4$ 1×2^2
                     2
                   $- 2$ 1×2^1
                     0   (frames 38–41)

3. (a) 11000   (frames 29–32 and 42–46)
    (b) 110001 (frames 47–54)
    (c) 1011001010   110011
                        x 1110
                           1100110
                          110011
                          100110010
                          110011
                          1011001010   (frames 55–60)

CHAPTER TWO
Octal and Hexadecimal Number Systems

People who work with computers a great deal find an elementary knowledge of the octal (base eight) and hexadecimal (base sixteen) number systems essential. Computer data on locations of values in their "memories" is usually available to programmers only in octal or hexadecimal numbers. Therefore, the ability to at least convert these numbers to decimal or binary is important. The hexadecimal number system is treated here in more detail than the octal number system. With the ability to add and subtract in both binary and hexadecimal, you would have no trouble expanding your skills to the octal number system.

When you complete this chapter, you will be able to:

- convert binary numbers to the octal number system;
- convert octal numbers to the decimal number system;
- convert decimal numbers to the octal number system;
- convert octal numbers to the binary number system;
- convert binary numbers to the hexadecimal number system;
- convert hexadecimal numbers to the binary or decimal number systems;
- add any two hexadecimal numbers;
- perform hexadecimal subtractions.

THE OCTAL NUMBER SYSTEM

The octal number system is used primarily to simplify working with the binary system. As you have discovered, binary numbers require long strings of ones and zeros that are difficult to keep track of. Conversions from binary to octal are very simple and can be done directly with no multiplication or subtraction.

1. The octal number system is based on powers of eight. How many different digits does it use? _____

- - - - - - - - - - - - - - -

8 (0, 1, 2, 3, 4, 5, 6, 7)

2. In the octal system, the value of each position in a number increases by powers of eight. Fill in the missing powers of eight in the table below.

8^0	8^1	8^2	8^3	8^4
	8		512	

- - - - - - - - - - - - - - - -

8^0	8^1	8^2	8^3	8^4
1	8	64	512	4096

3. The values of all powers of eight are equal to certain values of powers of two. For example, $2^3 = 8^1 = 8$. Complete the following chart.

Power of Two	Power of Eight	Decimal Value
	8^0	
2^3	8^1	8
	8^2	
	8^3	

- - - - - - - - - - - - - - - -

Power of Two	Power of Eight	Decimal Value
2^0	8^0	1
2^3	8^1	8
2^6	8^2	64
2^9	8^3	512

4. As shown in the chart above, the powers of two increase by three
   each time the powers of eight increase by one. Three positions in
   a binary number correspond to one position in an octal number.
   Any three-digit binary number can be replaced by a one-digit octal
   number. For conversion to octal, a binary number of less than
   three digits is padded with leading zeros, which, of course, have
   no effect on its value.

   The chart below gives the first ten counting numbers in the binary, octal, and decimal number systems. Notice that every three-
   digit number under binary has a one-digit octal equivalent.

Binary	Octal	Decimal
000	0	0
001	1	1
010	2	2
011	3	3
100	4	4
101	5	5
110	6	6
111	7	7
1000	10	8
1001	11	9

Write the octal equivalent of each of the following binary numbers. Use the chart for reference.

(a) 100 _____   (b) 101 _____

(c) 000 _____   (d) 111 _____

- - - - - - - - - - - - - - -

(a) 4; (b) 5; (c) 0; (d) 7

5. Octal numbers are converted to binary by replacing each octal digit with the three binary digits that correspond with it. For example, octal 24 is binary 010 100, or 10100.

Write the binary equivalent of each of the following octal numbers. Add leading zeros if needed.

(a) 367 _____

(b) 105 _____

(c) 2436 _____

- - - - - - - - - - - - - - - -

(a) 011 110 111 or 11110111; (b) 001 000 101 or 1000101;
(c) 010 100 011 110 or 10100011110

6. Converting from binary to octal is almost as easy as converting from octal to binary. First divide the binary string into groups of three digits, starting at the right (or at the binary point). Add leading zeros to the leftmost group until there are three digits. Then substitute the appropriate octal digit for each group of three binary digits.

The binary number below is already divided into groups of three. The leftmost group is padded with leading zeros. Refer to the chart in frame 4 and write the octal equivalent of this number.

$$001\ 111\ 001\ 011\ 101$$

_____

- - - - - - - - - - - - - -

17135_8

7. Divide the following binary numbers into groups of three. Add leading zeros if needed. Then write the octal equivalents.

(a) 111000101 _____

(b) 101010 _____

(c) 11001010 _____

- - - - - - - - - - - - - -

(a) 705; (b) 52; (c) 312

OCTAL–DECIMAL CONVERSION

8. Octal numbers are converted to decimal numbers just as binary numbers are, but using the values of powers of eight. In the example below, the octal 367 is converted to a decimal number.

$$3 \quad 6 \quad 7_8$$

$$7 \times 8^0 = \quad 7 \ (7 \times 1)$$
$$6 \times 8^1 = \quad 48 \ (6 \times 8)$$
$$3 \times 8^2 = \underline{192} \ (3 \times 64)$$

$$247 \quad \text{decimal equivalent}$$

Use this method to find the decimal equivalent of 240_8.

- - - - - - - - - - - - - -

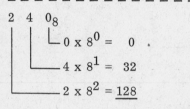

$$2 \quad 4 \quad 0_8$$

$$0 \times 8^0 = \quad 0 \ .$$
$$4 \times 8^1 = \quad 32$$
$$2 \times 8^2 = \underline{128}$$

$$160 \quad \text{decimal equivalent}$$

9. Find the decimal equivalent of octal 1601.

- - - - - - - - - - - - - -

897

DECIMAL-OCTAL CONVERSION

10. Decimal-octal conversion uses the same method as decimal-binary conversion, except the values of the powers of eight are subtracted. Any value may be subtracted up to seven times. (Seven is one less than eight. In binary a value was subtracted only once. One is one less than two, the base in question.)

   Below is an example of the conversion of decimal 777 to an octal number.

$$
\begin{array}{rl}
777 & \\
- 512 & \text{(value of } 8^3) \\
\hline
265 & \\
- 64 & (8^2) \\
\hline
201 & \\
- 64 & (8^2) \\
\hline
137 & \\
- 64 & (8^2) \\
\hline
73 & \\
- 64 & (8^2) \\
\hline
9 & \\
- 8 & (8^1) \\
\hline
1 & \\
- 1 & (8^0) \\
\hline
0 & \\
\end{array}
$$

The octal equivalent of decimal 777 is written 1411_8. 8^2 was subtracted four times, so a 4 appears in the 8^2 position. 8^3, 8^1, and 8^0 were each subtracted once, so a 1 is written in each of these positions.

   Use this method to convert decimal 198 to an octal number.

- - - - - - - - - - - - - -

306_8 (You subtracted 8^2 or 64 three times and 8^0 or 1 six times.)

11. For relatively large numbers, you may have to calculate some higher powers of eight. Can you convert decimal 10,000 to its octal equivalent?

- - - - - - - - - - - - - -

23420_8 (You subtracted 8^4 or 4096 two times, 8^3 or 512 three times, 8^2 or 64 four times, and 8^1 or 8 two times.)

THE HEXADECIMAL NUMBER SYSTEM

The hexadecimal, or base sixteen, number system is familiar to many programmers, since some computers produce data in this number system. It is very useful for the person who expects to work with computers to understand the basics of hexadecimal numbering, as well as addition and subtraction in this number system.

12. The hexadecimal number system uses the base sixteen. How would you read 12_{16}? _____

- - - - - - - - - - - - - -

"one-two, base sixteen" (not "twelve, base sixteen")

13. In the hexadecimal number system, sixteen symbols are used for values from zero to fifteen. They are, in order, 0, 1, 2, 3, 4, 5, 6, 7, 8, 9, A, B, C, D, E, and F.
    Write the hexadecimal digit that is equivalent to the following decimal numbers.

(a) 9_{10} _____

(b) 13_{10} _____

(c) 10_{10} _____

- - - - - - - - - - - - - - -

(a) 9; (b) D; (c) A

(The actual hexadecimal symbols used for decimal ten through fifteen are arbitrary. The ones shown here are used fairly often.)

14. The chart below gives decimal, hexadecimal, and binary equivalents
for the first thirty-two numbers in each system.

DECIMAL, HEXADECIMAL, AND BINARY NOTATION

Decimal	Hexa-decimal	Binary	Decimal	Hexa-decimal	Binary
0	0	0000	16	10	10000
1	1	0001	17	11	10001
2	2	0010	18	12	10010
3	3	0011	19	13	10011
4	4	0100	20	14	10100
5	5	0101	21	15	10101
6	6	0110	22	16	10110
7	7	0111	23	17	10111
8	8	1000	24	18	11000
9	9	1001	25	19	11001
10	A	1010	26	1A	11010
11	B	1011	27	1B	11011
12	C	1100	28	1C	11100
13	D	1101	29	1D	11101
14	E	1110	30	1E	11110
15	F	1111	31	1F	11111

Give the decimal equivalent of each of the following numbers.

(a) 1111_2 _____

(b) 11_{16} _____

(c) 20_{16} _____

(d) $1B_{16}$ _____

(e) 10000_2 _____

- - - - - - - - - - - - - - -

(a) 15; (b) 17; (c) 32; (d) 27; (e) 16

15. The hexadecimal number system uses base _____.

- - - - - - - - - - - - - - -

sixteen

16. The hexadecimal system is based on powers of sixteen. It will be useful to learn the values of a few of these powers. Complete the table below.

16^3	16^2	16^1	16^0	16^{-1}	16^{-2}	16^{-3}
				$\frac{1}{16}$		

— — — — — — — — — — — — — —

16^3	16^2	16^1	16^0	16^{-1}	16^{-2}	16^{-3}
4096	256	16	1	$\frac{1}{16}$	$\frac{1}{256}$	$\frac{1}{4096}$

17. Do you remember the values of some of the powers of two? $4096 = 2^{12}$ and $256 = 2^8$ are the same as the values of 16^3 and 16^2. The value of 16^1 is the same as which power of two? _____

— — — — — — — — — — — — — —

2^4

BINARY-HEXADECIMAL CONVERSION

18. In our study of octal numbers we saw that the value of 8^1 is the same as 2^3. We also found that three binary digits can be replaced by one octal digit. Now we find that the value of 16^1 is the same as 2^4. This indicates that you could substitute one hexadecimal digit for:

___(a) sixteen binary digits
___(b) three octal digits
___(c) one octal digit
___(d) four binary digits

— — — — — — — — — — — — — —

(d)

19. One hexadecimal digit can be directly substituted for four binary digits. Divide the binary number into groups of four, working left and right from the binary point.

   Divide the following binary numbers into appropriate groups for conversion to hexadecimal. Add leading or following zeros if necessary.

   (a) 100111.1101_2 _____

   (b) 1101110.111_2 _____

   (c) 100110110_2 _____

   - - - - - - - - - - - - - - - -

   (a) $0010\ 0111\ .1101_2$; (b) $0110\ 1110\ .1110_2$; (c) $0001\ 0011\ 0110_2$

20. Write the hexadecimal equivalent of the three binary numbers you divided in the last frame. Refer to the chart in frame 14 if necessary.

   (a) _____

   (b) _____

   (c) _____

   - - - - - - - - - - - - - - - -

   (a) $27.D_{16}$ ("two-seven-point-D, base sixteen")
   (b) $6E.E_{16}$ ("six-E-point-E, base sixteen")
   (c) 136_{16} ("one-three-six, base sixteen")

21. Convert the binary numbers below to the hexadecimal number system.

   (a) 111011111001.11101_2 _____

   (b) 1010101110000_2 _____

   (c) 1111111.1111_2 _____

   - - - - - - - - - - - - - - - -

   (a) $1110\ 1111\ 1001\ .1110\ 1000_2 = EF9.E8_{16}$
   (b) $0001\ 0101\ 0111\ 0000_2 = 1570_{16}$
   (c) $0111\ 1111\ .1111_2 = 7F.F_{16}$

22. Conversion from hexadecimal to the binary number system is done in reverse. For each hexadecimal digit you substitute

   _____.

_ _ _ _ _ _ _ _ _ _ _ _ _ _

four binary digits

23. Convert the hexadecimal numbers below to their binary equivalents.

(a) ABC_{16} _____

(b) 3.7_{16} _____

(c) $16.DE_{16}$ _____

_ _ _ _ _ _ _ _ _ _ _ _ _ _

(a) 101010111100; (b) 0011.0111 (or 11.0111);
(c) 00010110.11011110 (or 10110.1101111)
(Leading and following zeros may be included or omitted.)

HEXADECIMAL-DECIMAL CONVERSION

24. The chart below gives decimal values of some of the powers of sixteen.

16^3	16^2	16^1	16^0	16^{-1}	16^{-2}
4096	256	16	1	$\dfrac{1}{16}$	$\dfrac{1}{256}$

This chart can be used to find the decimal equivalents of hexadecimal numbers. For example:

$$3E7_{16} = (3 \times 16^2) + (14 \times 16^1) + (7 \times 16^0)$$

$$= (3 \times 256) + (14 \times 16) + (7 \times 1)$$

$$= 768 + 224 + 7$$

$$= 999$$

$$3E7_{16} = 999_{10}$$

The decimal equivalent of each hexadecimal digit is used in performing the multiplication.

Find the decimal equivalent of the hexadecimal number below.

(a) $104F_{16}$ = (___ x 16^3) + (___ x 16^2) + (___ x 16^1) + (___ x 16^0)

(b)   = (___ x ___) + (___ x 256) + (___ x ___) + (___ x ___)

(c)   = _____ + _____ + _____ + _____

(d) $104F_{16}$ = _____

- - - - - - - - - - - - -

(a) $104F_{16}$ = (1 x 16^3) + (0 x 16^2) + (4 x 16^1) + (15 x 16^0)
(b)   = (1 x 4096) + (0 x 256) + (4 x 16) + (15 x 1)
(c)   = 4096 + 0 + 64 + 15
(d) $104F_{16}$ = 4175_{10}

25. Digits to the right of a hexadecimal point represent reciprocals of powers of sixteen, or negative exponents of sixteen. Thus:

$$.1_{16} = \frac{1}{16} = 1 \text{ x } 16^{-1}$$

$$.01_{16} = \frac{1}{256} = 1 \text{ x } 16^{-2}$$

Hexadecimal numbers that include a point are converted to decimal as shown below:

$$3.2_{16} = (3 \text{ x } 16^0) + (2 \text{ x } 16^{-1})$$

$$= (3 \text{ x } 1) + (2 \text{ x } \frac{1}{16})$$

$$= 3 + \frac{2}{16}$$

$$3.2_{16} = 3\frac{1}{8}$$

Convert the hexadecimal number below to a decimal.

(a) $1E.41_{16}$ = (___ x ___) + (___ x ___) + (___ x ___) + (___ x ___)

(b)   = (___ x ___) + (___ x ___) + (___ x ___) + (___ x ___)

(c)   = _____ + _____ + _____ + _____

(d) $1E.41_{16}$ = _____

- - - - - - - - - - - - - -

(a) $1E.41_{16}$ = (1 x 16^1) + (14 x 16^0) + (4 x 16^{-1}) + (1 x 16^{-2})

(b)   = (1 x 16) + (14 x 1) + (4 x $\frac{1}{16}$) + (1 x $\frac{1}{256}$)

(c)   = 16 + 14 + $\frac{4}{16}$ + $\frac{1}{256}$

(d) $1E.41_{16}$ = $30\frac{65}{256}$ (converting to the lowest common denominator, $\frac{1}{16} = \frac{16}{256}$)

26. Use the following format to find the decimal equivalent of 1798_{16}.

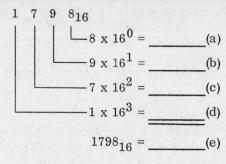

$$1 \quad 7 \quad 9 \quad 8_{16}$$

$$8 \times 16^0 = \underline{\qquad} \text{(a)}$$

$$9 \times 16^1 = \underline{\qquad} \text{(b)}$$

$$7 \times 16^2 = \underline{\qquad} \text{(c)}$$

$$1 \times 16^3 = \underline{\underline{\qquad}} \text{(d)}$$

$$1798_{16} = \underline{\qquad} \text{(e)}$$

- - - - - - - - - - - - - -

(a) 8; (b) 144; (c) 1792; (d) 4096; (e) 6040

27. What is the decimal equivalent of $243F_{16}$?

- - - - - - - - - - - - - -

$$2 \quad 4 \quad 3 \quad F_{16}$$

$$15 \times 16^0 = \quad 15$$

$$3 \times 16^1 = \quad 48$$

$$4 \times 16^2 = 1024$$

$$2 \times 16^3 = \underline{8192}$$

$$243F_{16} = 9279_{10}$$

HEXADECIMAL ADDITION

Addition in the hexadecimal number system follows the same rules as addition in decimal or binary. But working with non-numeric symbols makes it seem strange at first. In hexadecimal, just as in decimal, $4 + 5 = 9$. But, $7_{16} + 8_{16} = F_{16}$, not 15 as in decimal.

28. The hexadecimal addition table below gives the results of all additions of two one-digit numbers in the hexadecimal number system. You will learn to use this table. You are not expected to memorize it!

Hexadecimal Addition Table

+	1	2	3	4	5	6	7	8	9	A	B	C	D	E	F	10
1	02	03	04	05	06	07	08	09	0A	0B	0C	0D	0E	0F	10	11
2	03	04	05	06	07	08	09	0A	0B	0C	0D	0E	0F	10	11	12
3	04	05	06	07	08	09	0A	0B	0C	0D	0E	0F	10	11	12	13
4	05	06	07	08	09	0A	0B	0C	0D	0E	0F	10	11	12	13	14
5	06	07	08	09	0A	0B	0C	0D	0E	0F	10	11	12	13	14	15
6	07	08	09	0A	0B	0C	0D	0E	0F	10	11	12	13	14	15	16
7	08	09	0A	0B	0C	0D	0E	0F	10	11	12	13	14	15	16	17
8	09	0A	0B	0C	0D	0E	0F	10	11	12	13	14	15	16	17	18
9	0A	0B	0C	0D	0E	0F	10	11	12	13	14	15	16	17	18	19
A	0B	0C	0D	0E	0F	10	11	12	13	14	15	16	17	18	19	1A
B	0C	0D	0E	0F	10	11	12	13	14	15	16	17	18	19	1A	1B
C	0D	0E	0F	10	11	12	13	14	15	16	17	18	19	1A	1B	1C
D	0E	0F	10	11	12	13	14	15	16	17	18	19	1A	1B	1C	1D
E	0F	10	11	12	13	14	15	16	17	18	19	1A	1B	1C	1D	1E
F	10	11	12	13	14	15	16	17	18	19	1A	1B	1C	1D	1E	1F
10	11	12	13	14	15	16	17	18	19	1A	1B	1C	1D	1E	1F	20

The hexadecimal addition table is quite simple to use. To add 7 + 9, for example, you find 7 on the left and 9 on the top (or vice versa) and follow the row and column to their intersection. In the case of 7 + 9, it is at 10. Remember that this number is read "one-zero, base sixteen."

Use the hexadecimal addition table to find the following sums.

(a) E + E = _____

(b) 4 + C = _____

(c) 8 + 4 = _____

- - - - - - - - - - - - - - - - -

(a) 1C; (b) 10; (c) 0C, or C (Remember, leading zeros can be omitted.)

29. When adding hexadecimal numbers of more than one digit, proceed as in decimal addition, working from the right. If you have to carry a one, add it to the smaller of the next two numbers to be added. Then add as usual. For example, 3E + 12:

$$
\begin{array}{r}
3\ E \\
+\ \overset{+1}{1}\ 2 \\
\hline
5\ 0
\end{array}
$$

E + 2 = 10. Write down the 0, and carry the 1. Add the 1 to the smaller of the next two numbers to be added. Then add as usual: 3 + 2 = 5. The sum is 50_{16}.

Now find the following sums. Use the hexadecimal addition table on page 43.

(a)
$$
\begin{array}{r}
8\ F\ 9\ 7 \\
+\ D\ 4\ 4\ C \\
\hline
\end{array}
$$

(b)
$$
\begin{array}{r}
4\ 1\ .\ C\ 3 \\
+\ \ A\ .\ 5\ \ 0 \\
\hline
\end{array}
$$

- - - - - - - - - - - - - - -

(a)
$$
\begin{array}{r}
\overset{+1}{8}\ F\ 9\ 7 \\
+\ D\ 4\ 4\overset{+1}{}\ C \\
\hline
1\ 6\ 3\ E\ 3
\end{array}
$$

(b)
$$
\begin{array}{r}
4\ 1\overset{+1}{}\ .\ C\ 3 \\
+\ \ A\ .\ 5\ \ 0 \\
\hline
4\ C\ .\ 1\ 3
\end{array}
$$

30. Find these hexadecimal sums.

(a)
$$
\begin{array}{r}
6\ A\ E \\
+\ 1\ F\ A \\
\hline
\end{array}
$$

(b)
$$
\begin{array}{r}
4\ 5\ 2\ 8 \\
+\ 9\ 6\ 5\ 4 \\
\hline
\end{array}
$$

- - - - - - - - - - - - - - -

(a) 8A8; (b) DB7C

HEXADECIMAL SUBTRACTION

Hexadecimal subtraction, like binary subtraction, can be done by complementing. Here, though, we shall use our hexadecimal addition table and cope with the "borrowing" problem as it arises. Computers do use a complementing method for this operation; it is efficient for them, not for us.

31. Refer again to the hexadecimal addition table on page 43 as we work the problem $F_{16} - 9_{16}$ as an example.

F_{16} (minuend)
$- 9_{16}$ (subtrahend)
—————
?

Locate the column heading (across the top) of the digit in the subtrahend (here 9). Follow this column down until you find the minuend (here F). The heading of the row on the left that contains the minuend is the difference (here 6).

So $F_{16} - 9_{16} = 6_{16}$ in the hexadecimal number system. Remember that you take the number on the <u>bottom</u> of the subtraction problem and find it at the <u>top</u> of the table; then work from there.

Now try the following hexadecimal subtraction problem.

C
$-$ 4
———

(a) The minuend is _____ .

(b) The subtrahend is _____ .

(c) What column heading do you begin with? _____

(d) C − 4 = _____ .

— — — — — — — — — — — — — —

(a) C; (b) 4; (c) 4 (or the subtrahend); (d) 8

32. Find the following differences. Remember to take the number on the bottom and find it on the top of the hexadecimal addition table.

(a)   18
     $-$ A

(b)   12
     $-$ F

(c)   1D
     $-$ E

— — — — — — — — — — — — — —

(a) E; (b) 3; (c) F

33. The following hexadecimal subtraction problem is solved one column at a time, as in decimal subtraction. As in decimal, a number minus itself equals zero. Write the difference.

F8
$-$ A8
———

— — — — — — — — — — — — —

F8
$-$ A8
———
50

34. Find these differences.

  (a)   1F   (b)   107
         - 3   - 17

– – – – – – – – – – – – –

(a) 1C; (b) F0

35. Hexadecimal subtraction follows the same general rules as decimal subtraction. In decimal subtraction a borrowed 1 represents 10.

  In hexadecimal a borrowed 1 represents _____.

– – – – – – – – – – – – –

sixteen

36. The following hexadecimal subtraction problem calls for borrowing.

                    7A
                  - 1F

When you find F across the top of the hexadecimal addition table, there is no A in the column beneath it. But there is a 1A. So we borrow a 1 from the next digit over:

                    ⁶ ¹ᴬ
                    ⁷A̶
                  - 1F

Now we can find, using the hexadecimal addition table, that 1A − F = B, and 6 − 1 = 5. So the solution here is 5B.

                    ⁶ ¹ᴬ
                    ⁷A̶
                  - 1F
                    5B

     Rewrite the following problem to show a 1 borrowed from the C. Then use the hexadecimal addition table to find the difference.

                    C8
                  - AA

– – – – – – – – – – – – –

    ᴮ ¹⁸
    C̶8̶
  - AA
    1E

37. The point in a hexadecimal number does not affect the addition or subtraction operations, except that, as in decimal arithmetic, they must be lined up. Perform the following subtractions, rewriting if necessary.

(a)   F9D.5   (b) 9D.5 − 5.AF = _____
        − EB6.3

- - - - - - - - - - - -

(a) $\overset{E}{\cancel{F}}\overset{19}{\cancel{9}}$D.5   (b) $9\overset{C}{\cancel{D}}.\overset{14}{\cancel{5}}\overset{10}{\cancel{0}}$
        − EB6.3   − 5.AF
          E7.2   97.A1

38. All hexadecimal subtractions, like decimal subtractions, can be checked by adding the difference to the subtrahend. Check the results for the two problems in frame 37.

(a)   (b)

- - - - - - - - - - - -

(a)   E7.2   (b)   97.A1
       + EB6.3   + 5.AF
        F9D.5   9D.50

By now, you are probably reasonably competent in the use of the hexadecimal addition table for both addition and subtraction. You can convert hexadecimal numbers into binary and decimal. You can convert binary numbers into either octal (using groupings of three) or hexadecimal (using groupings of four). Now take the Self-Test on the following page.

SELF-TEST

This Self-Test will help you evaluate whether or not you have mastered the chapter objectives and are ready to go on to the next chapter. Answer each question to the best of your ability. Refer to the decimal, hexadecimal, and binary notation chart on page 37 and the hexadecimal addition table on page 43 when necessary. Correct answers are given at the end of the test.

1. Convert 101100111.1111001_2 to:

    (a) the octal number system

    (b) the hexadecimal number system

2. Convert each of the following numbers to the binary number system.

    (a) 417.02_8 _____

    (b) $F460_{16}$ _____

3. Give the decimal equivalent of each of the following numbers.

    (a) 417.1_8

    (b) $16EA_{16}$

    (c) $.1_{16}$

4. Use the hexadecimal addition table on page 43 to solve the following problems.

(a)   8 F B 6
     +C D 9 6

(b)   A B C D . 0 9
    + 1 2 3 4 . 5 6

(c)   CCC6
    − 8FB5

(d)   F8. 65
    − 9A. B0

ANSWERS TO SELF-TEST

Compare your answers to the Self-Test with the correct answers given below. If all of your answers are correct, you are ready to go on to the next chapter. If you missed any questions, study the frames indicated in parentheses following the answer.

1. (a) 547.744_8 (101 100 111 .111 100 100$_2$) (frames 4-5)
    (b) $167.F2_{16}$ (0001 0110 0111 .1111 0010$_2$) (frames 18-21)

2. (a) 100 001 111 .000 010$_2$ (frames 6-7)
    (b) 1111 0100 0110 0000$_2$ (frames 22-23)

3. (a) $271\dfrac{1}{8}$ $4\ \ 1\ \ 7\ .1_8$

$$1 \times 8^{-1} = \frac{1}{8}$$
$$7 \times 8^{0} = 7$$
$$1 \times 8^{1} = 8$$
$$4 \times 8^{2} = \underline{256}$$
$$271\frac{1}{8} \quad \text{(frames 8-9)}$$

(b) 5866 $1\ \ 6\ \ E\ \ A_{16}$

$$10 \times 16^{0} = 10$$
$$14 \times 16^{1} = 224$$
$$6 \times 16^{2} = 1536$$
$$1 \times 16^{3} = \underline{4096}$$
$$5866 \quad \text{(frames 24-27)}$$

(c) $\dfrac{1}{16}$ $.1_{16} = 1 \times 16^{-1} = \dfrac{1}{16^{1}} = \dfrac{1}{16}$ (frame 25)

4. (a) $\overset{+1}{8}$ F B 6
        $+$C $\overset{+1}{D}$ 9 6
        1 5 D 4 B (frames 28-30)

(b)   A B C D . 0 9
    $+$ 1 $\overset{+1}{2}\overset{+1}{3}$ 4 . 5 6
    B E 0 1 . 5 F (frames 28-30)

(c) $\overset{B}{\cancel{C}}\overset{16}{\cancel{C}}$C6
    $-$ 8FB5
    3D 11 (frames 31-37)

(d) $\overset{17}{E\ \overset{7}{\cancel{8}}}\ \overset{16}{\cancel{8}}$
    $\cancel{8}$8 . $\cancel{8}$5
    $-$ 9A . B0
    5D . B5 (frames 31-37)

CHAPTER THREE
Logic for Computers

Digital computers make decisions based on logic. The logic, like binary numbers, works efficiently with the two stable conditions of electronic circuits. In the binary number system, the conducting (closed or on) and nonconducting (open or off) states of each circuit are represented by 1 and 0. In logic, the two states are represented by T (true) and F (false).

In our discussion of logic, we will not study the electronic circuits themselves. The logic came before computers and can be thoroughly understood without them. The symbolic logic you learn here, however, underlies the problem-solving activities of the computer. A thorough understanding of logic will help you program more effectively and efficiently.

When you complete this chapter, you will be able to:

- identify a proposition;

- give the truth value of a proposition and its negation;

- give the truth value of a conjunction, given the values of its propositions;

- give the truth value of a disjunction, given the values of its propositions;

- give the truth value of an implication, given the values of its propositions;

- give the converse and contrapositive of an implication;

- recognize and use correctly the following symbols: $\sim$, $\wedge$, $\vee$, $\rightarrow$, $\leftrightarrow$, and $\equiv$;

- complete or construct truth tables to verify equivalences;

- identify DeMorgan's rules of logic.

Mathematical prerequisites for this chapter are minimal. You will use basic arithmetic and should recognize and be able to work with the following symbols:

$=$ equal to

$\neq$ not equal to

$<$ less than

$\nless$ not less than

$>$ greater than

$\ngtr$ not greater than

STATEMENTS AND PROPOSITIONS

1. A <u>statement</u> is any sentence that asserts or denies a property about some specific object or class of objects. Which of these sentences is a statement?

___(a) Are you happy?
___(b) I am six feet tall.
___(c) Dogs do not have five legs.

— — — — — — — — — — — — — — —

(b); (c) ("Are you happy?" does not assert or deny a specific property.)

2. An expression of equality or inequality is a mathematical sentence. The equality $3 + 6 = 9$ is a sentence which is also a statement. It asserts that the sum of 3 and 6 is equal to 9. Which of the following is a statement?

___(a) $3 - 1 = 4$
___(b) Children are good.
___(c) $7 \times 4 = 28$

— — — — — — — — — — — — — —

(a); (b); (c) ("$3 - 1 = 4$" is not true, and "children are good" is debatable, but each asserts a specific property. An expression such as $x + y = z$ is not a statement, however, until specific values are given to x, y, and z.)

3. A <u>proposition</u> is a statement which is either true or false. A statement such as "children are good" is not a proposition, since some people would say it is true and some would say it is false. A proposition is either true or false, never ambiguous. Which of the following is a proposition?

___(a) Some dogs are brown.
___(b) 2 + 7 = 8
___(c) 7 > 2 (7 is greater than 2)
___(d) Physics problems are very difficult.

– – – – – – – – – – – – – – –

(a); (b); (c) ("Physics problems are very difficult" is a statement, but not a proposition. Some students find physics problems easy.)

4. Every proposition is either true or false. Write T or F in front of each proposition below.

___(a) 7 x 8 = 64
___(b) Every statement is a proposition.
___(c) Every proposition is a statement.

– – – – – – – – – – – – – – –

(a) F; (b) F; (c) T

5. Match each sentence below with all appropriate terms.

___(a) The sun revolves around the moon.   1. Not a statement
___(b) 4 x 3 = 7   2. A statement, but not a proposition
___(c) Barbequed squid is tasty.   3. A proposition
___(d) What time is it?   4. True
   5. False

– – – – – – – – – – – – – – –

(a) 3, 5; (b) 3, 5; (c) 2 (ambiguous, neither true or false); (d) 1 (doesn't assert or deny a property)

A mathematical "open sentence," such as 3 + x = 7, is not a statement until x is given a specific value, as noted earlier. After a variable, such as x, is given a specific value, an equality or inequality containing it may be a statement, and a proposition.

NEGATION OF PROPOSITIONS

The negation of a proposition is its negative. If we use a small letter to represent a proposition, such as p, then we represent the negation of that proposition with ~p (read "not p").

6. When a proposition is true, its negation is false. When a proposition is false, its negation is true. "The earth revolves around the sun" is a proposition. Its negation could be stated as "the earth does not revolve around the sun" or "it is not true that the earth revolves around the sun." Is this negation true or false? _____

    _ _ _ _ _ _ _ _ _ _ _ _ _ _

    false

7. True or false is the <u>truth value</u> of a proposition. What is the truth value of the negation of the proposition, "The sun revolves around Mars"? _____

    _ _ _ _ _ _ _ _ _ _ _ _ _ _

    true

8. The truth value of propositions and their negations can be represented in a <u>truth table</u>. In the table below, p stands for a proposition and ~p stands for its negation. Complete the table by writing in truth values for ~p.

p	~p
T	
F	

_ _ _ _ _ _ _ _ _ _ _ _ _ _

p	~p
T	F
F	T

9. "Some birds can fly" is a proposition. Which of the following might be its negation?

___(a) All birds can fly.
___(b) No birds can fly.
___(c) Some birds cannot fly.
___(d) It is not true that some birds can fly.

– – – – – – – – – – – – – – – –

(b); (d)

10. Assume that "all dogs are black" is proposition p. Which of the following might be represented by ~p?

___(a) No dogs are black.
___(b) It is not true that all dogs are black.
___(c) Some dogs are not black.

– – – – – – – – – – – – – – –

(b); (c) ("No dogs are black" has the same truth value as p; therefore it cannot be ~p. "It is not true that all dogs are black" and "some dogs are not black" are different ways of stating the negation.)

11. Without changing the numbers, write a negation of 3 + 4 = 8.

_____

– – – – – – – – – – – – – –

$3 + 4 \neq 8$ or $3 + 4 < 8$ (Since 3 + 4 = 8 is false, you must write a proposition that is true.)

CONJUNCTION OF PROPOSITIONS

12. Propositions can be combined or joined in complex forms. For example, "some dogs are brown AND no dogs can fly" is a combination of two propositions. Any combination of two (or more) propositions joined by AND is a <u>conjunction</u>. We use the symbol $\wedge$, which is read "and." The conjunction of two propositions is represented by $p \wedge q$ and is read "p and q." The symbol $\wedge$ looks somewhat like the initial letter of AND.

"Some dogs are brown and no dogs can fly." Suppose we represent this conjunction as $p \wedge q$.

(a) The truth value of p is _____.

(b) The truth value of q is _____.

- - - - - - - - - - - - - -

(a) T; (b) T

13. A conjunction of two true propositions is always true. In p ∧ q, if
either p or q, or both, is false, then the conjunction is false. Complete the conjunction truth table below by writing T or F in each
space.

p	q	p ∧ q
T	T	
T	F	
F	T	
F	F	

- - - - - - - - - - - - - -

p	q	p ∧ q
T	T	T
T	F	F
F	T	F
F	F	F

(Refer to this truth table as necessary while
working through the next several frames.)

14. "All normal men have two legs and all men have three eyes" is a
conjunction. It is of the form p ∧ q. Fill in the following truth
values.

___(a) p
___(b) q
___(c) p ∧ q

- - - - - - - - - - - - - -

(a) T; (b) F; (c) F

15. "Dandelions are flowers and all robins are birds." What is the truth value of this conjunction? _____

– – – – – – – – – – – – – –

T   (Both p and q are true.)

16. Give the truth value of each conjunction below.

___(a) Birds have fur and fish have feathers.
___(b) Fish swim and fish have feathers.
___(c) Birds have feathers and squirrels have fur.
___(d) 2 + 3 = 5 and 2 x 3 = 4.

– – – – – – – – – – – – – –

(a) F; (b) F; (c) T; (d) F

17. "2 + 3 = 5 and 2 + 3 ≠ 5" is a conjunction and can be represented as p ∧ ~p. The truth value of p∧~p is always the same. It is _____.

– – – – – – – – – – – – – –

F   (p and ~p always have different truth values. Since one is always false, the truth value of p ∧ ~p is always F.)

18. A conjunction can have more than two propositions. The conjunction will be true only when all of its propositions are true. What is the truth value of the conjunction "1 + 2 = 3, and 2 + 3 = 5, and 3 + 5 = 8, and 5 + 8 = 12"? _____

– – – – – – – – – – – – – –

F   (One proposition is false, so the conjunction is false.)

19. Assume that p is true, q is false, and r is true. Give the truth value of each of the following conjunctions.

___(a) p ∧ ~r
___(b) q ∧ ~p
___(c) p ∧ q ∧ r
___(d) p ∧ ~q ∧ r
___(e) p ∧ q ∧ ~r

– – – – – – – – – – – – – –

(a) F (~r is false); (b) F (q and ~p are false); (c) F (q is false);
(d) T; (e) F (q is false)

DISJUNCTION OF PROPOSITIONS

20. Propositions can also be combined in <u>disjunctions</u>—that is, connected with OR. The symbol for the disjunction is ∨. We show the disjunction of two propositions with p ∨ q, and read it "p or q."

    A disjunction, two or more propositions joined by OR, is false only when <u>all</u> propositions in it are false. Which of the disjunctions below is false?

    ___(a) Birds fly or fish fly.
    ___(b) Elephants are fish or mice are birds.
    ___(c) Some dogs are brown or no dogs can fly.

    — — — — — — — — — — — — — — —

    (b) (In (a) one proposition is true; in (c) both are true. Therefore the disjunctions are true.)

21. A disjunction is true if only one of its propositions is true. Complete the disjunction truth table for two propositions below.

p	q	p ∨ q
T	T	
T	F	
F	T	
F	F	

— — — — — — — — — — — — — — —

p	q	p ∨ q
T	T	T
T	F	T
F	T	T
F	F	F

22. A disjunction p ∨ q is true if:

___(a) both p and q are true
___(b) p is true and q is false
___(c) p is false and q is true
___(d) both p and q are false

– – – – – – – – – – – – – –

(a); (b); (c)

23. "Some walls are brick or no floors are carpeted" is a disjunction of the form p ∨ q. Fill in the truth values below.

___(a) p
___(b) q
___(c) p ∨ q

– – – – – – – – – – – – – –

(a) T; (b) F; (c) T

24. What is the truth value of the disjunction "all birds are robins or no birds are orioles"? _____

– – – – – – – – – – – – – –

F (Both p and q are false.)

25. Give the truth value of each of the following disjunctions.

___(a) Birds have fur or fish have feathers.
___(b) Fish swim or fish have feathers.
___(c) Birds have feathers or squirrels have fur.
___(d) 2 + 3 = 5 or 2 x 3 = 4.

– – – – – – – – – – – – –

(a) F; (b) T; (c) T; (d) T

26. Some disjunctions of the form p ∨ ~p are given below.

"A kangaroo is a mammal or a kangaroo is not a mammal."
"$8^3 = 24$ or $8^3 \neq 24$."

The truth value of p ∨ ~p is always the same, no matter what the truth value of p. What is the truth value of p ∨ ~p? _____

– – – – – – – – – – – – –

T (Either p or ~p is always true; the other is always false.)

27. Write the symbol for each of the following.

    (a) The negation of p _____

    (b) The conjunction of p and q _____

    (c) The disjunction of p and q _____

- - - - - - - - - - - - - -

(a) $\sim p$; (b) $p \wedge q$; (c) $p \vee q$

28. Which of the following have truth value T?

    ___(a) $3 + 7 = 10$ or $9 - 2 = 11$.
    ___(b) $7 \times 3 = 73$ and $9 - 2 = 7$.
    ___(c) $8 + 4 \neq 12$ or $8 - 4 \neq 4$.
    ___(d) $2^2 = 4$ and $2^3 = 8$.

- - - - - - - - - - - - - -

(a); (d)

29. Assume that p is true and q is false. Give the truth value of each of the following.

    ___(a) $p \wedge \sim q$
    ___(b) $p \wedge q$
    ___(c) $p \vee \sim q$
    ___(d) $\sim p \vee q$
    ___(e) $\sim q \vee q$

- - - - - - - - - - - - - -

(a) T; (b) F; (c) T; (d) F; (e) T

30. Many propositions can be included in a disjunction, just as in a conjunction. The disjunction is false only when _every_ proposition in it is false.

    Give the truth value of each of the following disjunctions.

    ___(a) $3 + 6 = 9$, or $7 - 2 = 5$, or $9 \times 3 = 40$.
    ___(b) $9 - 7 = 0$, or $6 + 3 \neq 9$, or $1 + 1 = 1$, or $2 \times 7 = 15$.
    ___(c) $8^2 = 64$, or $8^3 = 512$, or $2^6 = 32$.

- - - - - - - - - - - - - -

(a) T; (b) F; (c) T

31. Parentheses are used to group logic symbols just as they are used
to group symbols in algebra. Operations within the parentheses are
performed first. You find the truth value of ~(p ∧ q), for example,
by first finding the truth value of p ∧ q, then negating it. Assume
that p is true and q is false.

(a) The truth value of p ∧ q is _____.

(b) The truth value of ~(p ∧ q) is _____.

_ _ _ _ _ _ _ _ _ _ _ _ _ _ _

(a) F; (b) T

32. The truth table below shows all the possible values of ~(p ∧ q) for
all values of p and q.

p	q	p ∧ q	~(p ∧ q)
T	T	T	F
T	F	F	T
F	T	F	T
F	F	F	T

Construct a similar truth table to show all the possible truth values
of ~(p ∨ q).

- - - - - - - - - - - -

p	q	p ∨ q	~(p ∨ q)
T	T	T	F
T	F	T	F
F	T	T	F
F	F	F	T

DeMORGAN'S RULES OF LOGIC

DeMorgan's rules of logic give simple equivalents for negations of conjunctions and disjunctions. These equivalents allow digital computers to process data more efficiently. You will learn DeMorgan's two simple rules in this section. One gives the equivalent of ~(p ∧ q); the other gives the equivalent of ~(p ∨ q).

33. One of DeMorgan's rules of logic gives the equivalent of ~(p ∧ q). With which of the examples below could this equivalent be used?

___(a) It is not true that all cats are mammals and all mice are insects.

___(b) It is false that Washington was a President or Lincoln was a President.

___(c) It is not the case that 6 + 1 = 7 and 9 − 2 = 7.

- - - - - - - - - - - - - -

(a); (c) (Both (a), which is true, and (c), which is false, can be represented as ~(p ∧ q).)

34. DeMorgan's rule is ~(p ∧ q) ≡ ~p ∨ ~q. The symbol ≡ represents equivalence. "It is not the case that 6 + 1 = 7 and 9 − 2 = 7" is equivalent to "6 + 1 ≠ 7 or 9 − 2 ≠ 7." Both the original proposition and its equivalent are false.

"It is not true that all cats are mammals and all mice are insects" is equivalent to which of the following?

___(a) All cats are mammals or all mice are insects.

___(b) All cats are not mammals or all mice are not insects.

___(c) Some cats are not mammals and some mice are not insects.

‒ ‒ ‒ ‒ ‒ ‒ ‒ ‒ ‒ ‒ ‒ ‒ ‒ ‒

(b)

35. Truth tables are useful in proving that certain logical expressions
have the same truth value in all cases. By this we mean they are
equivalent and we use the symbol ≡ to represent equivalence. You
have seen that one of DeMorgan's rules is ~(p ∧ q) ≡ ~p ∨ ~q. You
are going to show this is true by completing the truth table below.
The first part of the table (up to column A) you saw in frame 32.
After you complete this table, column A and column B will have the
same entries, showing that the expressions are equivalent. Now
complete the table.

| | | | A | | | B |
p	q	p ∧ q	~(p ∧ q)	~p	~q	~p ∨ ~q
T	T	T	F			
T	F	F	T			
F	T	F	T			
F	F	F	T			

‒ ‒ ‒ ‒ ‒ ‒ ‒ ‒ ‒ ‒ ‒ ‒ ‒ ‒

| | | | A | | | B |
p	q	p ∧ q	~(p ∧ q)	~p	~q	~p ∨ ~q
T	T	T	F	F	F	F
T	F	F	T	F	T	T
F	T	F	T	T	F	T
F	F	F	T	T	T	T

Column A has the same truth values as column B; therefore,
~(p ∧ q) ≡ ~p ∨ ~q.

36. The expression used in the last frame, ~(p ∧ q), is read "NOT (p and q)." In computer terms, this is often called a NAND, for NOT-AND.

    Some symbolic circuits are shown below. Two truth values come in (are input) and one truth value is the output.

Fill in the output truth values for the symbolic circuits below.

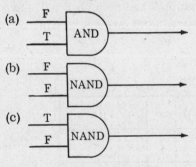

(a)

(b)

(c)

- - - - - - - - - - - - - - - -

(a) F [(F∧T)] ; (b) T [~(F ∧ F)]; (c) T [~(T ∧ F)]

37. You have seen that ~(p ∧ q) is a combination of NOT and AND. Which of the expressions below might be a combination of NOT and OR?

____(a) ~p ∨ ~q
____(b) ~(p ∨ q)
____(c) ∨ (p~q)

- - - - - - - - - - - - - - - -

(b) (Choice (a) was shown to be equivalent to NAND in frame 35.)

38. DeMorgan's second rule states ~(p ∨ q) ≡ ~p ∧ ~q. "It is not true that 6 + 1 = 7 or 9 − 2 = 7" is equivalent to:

____(a) 6 + 1 ≠ 7 or 9 − 2 ≠ 7.
____(b) 6 + 1 ≠ 7 and 9 − 2 ≠ 7.
____(c) 6 + 1 = 7 and 9 − 2 = 7.

- - - - - - - - - - - - - - - -

(b)

39. The negation of a disjunction, ~(p ∨ q), is sometimes called NOR, for NOT-OR. In frame 32 you constructed a truth table giving the values of ~(p ∨ q). DeMorgan's second rule is ~(p ∨ q) ≡ ~p ∧ ~q. Complete the truth table below to verify this rule. Columns A and B should be identical in truth values.

A ⏜   B ⏜

p	q	p ∨ q	~(p ∨ q)	~p	~q	~p ∧ ~q
T	T	T	F			
T	F	T	F			
F	T	T	F			
F	F	F	T			

- - - - - - - - - - - - - - - -

A ⏜   B ⏜

p	q	p ∨ q	~(p ∨ q)	~p	~q	~p ∧ ~q
T	T	T	F	F	F	F
T	F	T	F	F	T	F
F	T	T	F	T	F	F
F	F	F	T	T	T	T

40. NOR truth values can be found in the truth table you just completed, in either column A or column B. Refer to these when necessary as you fill in the output truth values for these OR and NOR symbolic circuits.

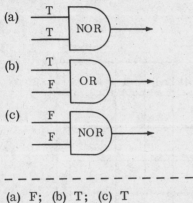

(a) T T NOR

(b) T F OR

(c) F F NOR

- - - - - - - - - - - - - -

(a) F; (b) T; (c) T

41. DeMorgan's rules give equivalent values for NAND and NOR expressions. One of DeMorgan's rules is $\sim(p \vee q) \equiv \sim p \wedge \sim q$. The other rule is:

___(a) $\sim(p \vee q) \equiv \sim p \vee \sim q$
___(b) $\sim(p \wedge q) \equiv \sim p \wedge \sim q$
___(c) $\sim(p \wedge q) \equiv \sim p \vee \sim q$

- - - - - - - - - - - - - -

(c)

42. Match the following expressions.

___(a) $\sim(p \vee q)$ 1. NAND
___(b) $\sim(p \wedge q)$ 2. NOR
___(c) $\sim p \vee \sim q$
___(d) $\sim p \wedge \sim q$

- - - - - - - - - - - - - -

(a) 2; (b) 1; (c) 1; (d) 2

43. Fill in output truth values for these symbolic circuits.

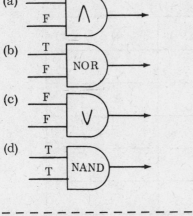

(a) T / F ∧
(b) T / F NOR
(c) F / F ∨
(d) T / T NAND

– – – – – – – – – – – – – –

(a) F; (b) F; (c) F; (d) F

44. Write the two equivalences that are called DeMorgan's rules of logic.

_____ and _____

– – – – – – – – – – – – – –

~(p ∧ q) ≡ ~p ∨ ~q; ~(p ∨ q) ≡ ~p ∧ ~q (in either order)

TRUTH TABLES

In the last section you used truth tables to prove equivalences by showing that two expressions had the same truth values. In this section we will do the same with a few more equivalences, and also with some rather complex expressions that combine ∧ and ∨ .

45. Truth values of complex expressions are found using truth tables. We have seen that an expression using two propositions, p and q, requires four lines to cover all possible truth values. An expression using three variables (say p, q, and r) requires eight lines in its truth table.

The partial table below shows all values of p, q, and r. Complete the table to show values of the expression (p ∨ q) ∧ (p ∨ r).

p	q	r	p ∨ q	p ∨ r	(p ∨ q) ∧ (p ∨ r)
T	T	T			
T	T	F			
T	F	T			
T	F	F			
F	T	T			
F	T	F			
F	F	T			
F	F	F			

- - - - - - - - - - - - - -

p	q	r	p ∨ q	p ∨ r	(p ∨ q) ∧ (p ∨ r)
T	T	T	T	T	T
T	T	F	T	T	T
T	F	T	T	T	T
T	F	F	T	T	T
F	T	T	T	T	T
F	T	F	T	F	F
F	F	T	F	T	F
F	F	F	F	F	F

46. You have found the truth values for (p ∨ q) ∧ (p ∨ r). In logic,
(p ∨ q) ∧ (p ∨ r) ≡ p ∨ (q ∧ r). If you find the truth values of
p ∨ (q ∧ r), you can compare the values to verify the equivalence.
   Complete the truth table below. Remember to find the values
of the expression in parentheses first.

p	q	r		
T	T	T		
T	T	F		
T	F	T		
T	F	F		
F	T	T		
F	T	F		
F	F	T		
F	F	F		

- - - - - - - - - - - - - - - - -

p	q	r	q ∧ r	p ∨ (q ∧ r)
T	T	T	T	T
T	T	F	F	T
T	F	T	F	T
T	F	F	F	T
F	T	T	T	T
F	T	F	F	F
F	F	T	F	F
F	F	F	F	F

(The final columns in this
table and in the table in
frame 45 are identical.
Therefore the equivalence
is valid.)

IMPLICATION OF PROPOSITIONS

In logic, an <u>implication</u> involves two propositions, one of which implies the other. There are two forms of logical implication. One is written p → q and is read "p implies q" or "if p, then q." The other is written p ↔ q and is read "p strictly implies q" or "if and only if p, then q." We will study the regular implication, "if p, then q" first.

We frequently make implication statements in our everyday speech. You may say "if I wash my car today, then it will rain tomorrow" or "if that politician is telling the truth, then I'm a monkey's uncle." In our speech, the propositions usually have some connection. In logic, the propositions need have no connection. We are concerned only with truth value. "If the sun revolves around the earth, then John Wilkes Booth was the eighteenth president" is a valid logical implication. As you will learn in the next section, an implication is false only when p is true and q is false.

47. An implication, p → q, is false only when p is true and q is false. Complete the implication truth table below.

p	q	p → q
T	T	
T	F	
F	T	
F	F	

- - - - - - - - - - - - - - -

p	q	p → q
T	T	T
T	F	F
F	T	T
F	F	T

48. The implication p → q is usually read "if p, then q." Refer to the truth table in frame 47 and give the truth value of each of the following implications.

___(a) If all robins are birds, then no monkey has a tail.
___(b) If $3^2 = 8$, then $2^3 = 6$.
___(c) If 3 + 4 = 8, then 5 + 9 = 14.
___(d) If some dogs have four legs, then Johnny Cash is a singer.

– – – – – – – – – – – – – –

(a) F (p is true, q is false); (b) T (p is false, q is false);
(c) T (p is false, q is true); (d) T (p is true, q is true)

49. If p is false, then implication p → q is always true. An implication beginning "if I am ten feet tall" will always be true. A false proposition implies anything at all. If q is true, the implication p → q is likewise always true. Any statement that ends, "then I am less than ten feet tall" will always be true. A true proposition is implied by any proposition at all. In what single case, then, would p → q be false? _____

– – – – – – – – – – – – – –

when p is true and q is false

50. Assume p and q are true propositions. Give the truth value of each of the following.

___(a) p ∨ q
___(b) p ∧ ~q
___(c) ~p → q
___(d) q → ~p

– – – – – – – – – – – – – –

(a) T; (b) F; (c) T; (d) F

51. Complete the truth table below to find all the truth values of
p → (q ∨ ~p). Remember that an implication is false only when
something true is used to imply something false.

p	q	~p	q ∨ ~p	p → (q ∨ ~p)
T	T	F		
T	F			
F	T			
F	F			

- - - - - - - - - - - -

p	q	~p	q ∨ ~p	p → (q ∨ ~p)
T	T	F	T	T
T	F	F	F	F
F	T	T	T	T
F	F	T	T	T

(In only one case does p imply a false proposition, when (q ∨ ~p) is
false.)

52. Complete the truth table below to show all truth values of
(p ∧ r) → q.   Remember to find the values of the expression in
parentheses first.

p	q	r	
T	T	T	
T	T	F	
T	F	T	
T	F	F	
F	T	T	
F	T	F	
F	F	T	
F	F	F	

- - - - - - - - - - - - - - -

p	q	r	p ∧ r	(p ∧ r) → q
T	T	T	T	T
T	T	F	F	T
T	F	T	T	F
T	F	F	F	T
F	T	T	F	T
F	T	F	F	T
F	F	T	F	T
F	F	F	F	T

53. The <u>contrapositive</u> of an implication p → q is written ∼q → ∼p. Consider the implication "6 + 1 = 7 → 1 < 2." The contrapositive of this is "1 ≮ 2 → 6 + 1 ≠ 7."

    "If all dogs are mammals, then all beagles are mammals." The contrapositive of this is "if all beagles are not mammals, then

    _____."

    - - - - - - - - - - - - - -

    all dogs are not mammals

54. For any implication, you can write its contrapositive. The contrapositive of p → q is ∼q → ∼p. An implication and its contrapositive are always equivalent. Complete the truth table below to show that the truth value of any implication is the same as the truth value of its contrapositive.

p	q	p → q	∼q	∼p	∼q → ∼p
T	T				
T	F				
F	T				
F	F				

- - - - - - - - - - - - - -

p	q	p → q	∼q	∼p	∼q → ∼p
T	T	T	F	F	T
T	F	F	T	F	F
F	T	T	F	T	T
F	F	T	T	T	T

A   B

(p → q ≡ ∼q → ∼p since the truth values in columns A and B are identical.)

55. "If 3 + 6 = 9, then 5 x 4 = 19" is an implication. Which of the follow-
    ing is its contrapositive?

    ___(a) If 5 x 4 = 19, then 3 + 6 = 9.
    ___(b) If 3 + 6 ≠ 9, then 5 x 4 ≠ 19.
    ___(c) If 5 x 4 ≠ 19, then 3 + 6 ≠ 9.

    - - - - - - - - - - - - - -

    (c) $(p \to q \equiv \sim q \to \sim p)$

56. The contrapositive of $(p \wedge q) \to \sim r$ is $r \to \sim(p \wedge q)$, or
    $\sim(\sim r) \to \sim(p \wedge q)$. (A double NOT can be either included or
    dropped.) "If 6 x 4 = 24, then it is not true that 2/3 x 6 = 4 and
    1/4 x 24 = 4." This could be written as "if it is not true that
    6 x 4 ≠ 24," and have the same meaning. Its contrapositive would
    be "if 2/3 x 6 = 4 and 1/4 x 24 = 4, then 6 x 4 ≠ 24."
    Write the contrapositive of each of the following.

    (a) $(p \vee q) \to r$ _____

    (b) $(\sim p) \to q$ _____

    (c) $p \to (q \wedge r)$ _____

    - - - - - - - - - - - - - -

    (a) $\sim r \to \sim(p \vee q)$; (b) $\sim q \to p$ [or $\sim q \to \sim(\sim p)$]; (c) $\sim(q \wedge r) \to \sim p$

57. The <u>converse</u> of an implication $p \to q$ is $q \to p$. "If 3 + 3 = 6, then
    3 = 1/2 x 6" has the converse "if 3 = 1/2 x 6, then 3 + 3 = 6."
    Write the converse of each of the following implications.

    (a) If I am a human, then I am not a dog. _____

    _____

    (b) If the sun rises in the west, then you are in the Southern Hemi-

    sphere. _____

    _____

    - - - - - - - - - - - - - -

    (a) If I am not a dog, then I am human. (The contrapositive is "if I
        am a dog, then I am not human.")
    (b) If you are in the Southern Hemisphere, then the sun rises in the
        west. (The contrapositive is "if you are not in the Southern
        Hemisphere, then the sun does not rise in the west.")

58. The converse of an implication p → q is simply q → p. Construct a truth table that will show that the converse of an implication is not equivalent to the implication.

p	q	
T	T	
T	F	
F	T	
F	F	

- - - - - - - - - - - - -

p	q	p → q	q → p
T	T	T	T
T	F	F	T
F	T	T	F
F	F	T	T

(The truth values of p → q and q → p are not the same; therefore they are not equivalent.)

59. Write the converse of each of the following.

(a) (p ∨ q) → r _____

(b) (~p) → q _____

(c) p → (q ∧ r) _____

- - - - - - - - - - - - -

(a) r → (p ∨ q); (b) q → ~p; (c) (q ∧ r) → p

60. "If 5 + 6 = 9, then 17 − 4 = 13" is an implication. Write the following:

(a) the contrapositive _____

(b) the converse _____

- - - - - - - - - - - - - -

(a) "If $17 - 4 \neq 13$, then $5 + 6 \neq 9$."

(b) "If $17 - 4 = 13$, then $5 + 6 = 9$."

61. Each of the three expressions in frame 60 has a truth value. Fill them in below.

___(a) implication

___(b) contrapositive

___(c) converse

- - - - - - - - - - - - - -

(a) T; (b) T; (c) F

STRICT IMPLICATION OF PROPOSITIONS

Many computer languages use an IF statement as a conditional branching device. The programmer may say to the computer (in effect), "IF the value of a certain variable is negative, THEN print some information." Or he may say, "IF you have already printed fifty lines, THEN start printing the next line on a new page." Statements like these are strict implications. In these IF statements, q will become true if and only if p is true. If p is not true, q will not be carried out. The programmer, of course, always provides an alternate action for the computer in case p is false. This strict implication is symbolized $p \leftrightarrow q$. We usually read it "if and only if p, then q."

62. The expression $p \leftrightarrow q$ is read "if and only if p, then q." When p is true, q is true; when p is false, q is false. In these two cases the strict implication, $p \leftrightarrow q$, can be said to be true. In all other cases, $p \leftrightarrow q$ is false. Complete the strict implication truth table below.

p	q	$p \leftrightarrow q$
T	T	
T	F	
F	T	
F	F	

- - - - - - - - - - - - - -

p	q	p ↔ q
T	T	T
T	F	F
F	T	F
F	F	T

63. Both of the strict implications below are true according to the truth table you just completed.

"If and only if 3 + 3 = 6, then 3 x 2 = 6."
"If and only if 3 + 2 = 6, then 6 − 2 = 3."

Refer to the table if necessary and give the truth value of each of the following.

___(a) If and only if 9 x 3 = 27, then 9^2 = 72.
___(b) If and only if fish have fur, then some birds fly.
___(c) If and only if some birds fly, they some birds have feathers.

- - - - - - - - - - - - - -

(a) F; (b) F; (c) T

64. The converse and contrapositive of strict implications are written just like those for regular implications, except the symbol ↔ is used. Write the converse and contrapositive for p ↔ q.

(a) converse _____

(b) contrapositive _____

- - - - - - - - - - - - - -

(a) q ↔ p; (b) ~q ↔ ~p

65. Complete the truth table below.

p	q	p ←→ q	q ←→ p
T	T		
T	F		
F	T		
F	F		

Is the converse of a strict implication equivalent to the strict impli-

cation? _____

----

p	q	p ←→ q	q ←→ p
T	T	T	T
T	F	F	F
F	T	F	F
F	F	T	T

Yes (The truth table shows that the converse of a strict implication
is equivalent to the strict implication.)

66. A strict implication in which p is true and q is false, or p is false
and q is true, is sometimes said to be invalid. Which of the following strict implications might be said to be valid?

___(a) If and only if 6 + 3 = 9, then 3 + 6 = 9.
___(b) If and only if fish swim, then people are fish.
___(c) If and only if 9 x 9 = 99, then 99 > 0.

(a)

67. IF statements in computer programs always use the strict implication. An IF statement may tell the computer, "If this person has earned less than $9000 this year, calculate an FICA deduction." In which of the cases below will an FICA deduction be calculated?

___(a) If and only if the person has earned less than $9000.
___(b) Always, since a true conclusion is implied by any proposition.
___(c) Never, since a false conclusion is implied by any proposition.
___(d) Jerry Jenkins has earned $7999.25 so far this year.

- - - - - - - - - - - - - - -

(a); (d)

68. Assume that p, q, and r are true propositions. Give the truth value of each of the following.

___(a) $\sim p \to q$
___(b) $(p \vee \sim r) \to \sim(q \wedge \sim p)$
___(c) $\sim q \leftrightarrow p$
___(d) $(p \vee \sim r) \leftrightarrow \sim(q \wedge \sim p)$

- - - - - - - - - - - - - - - - -

(a) T; (b) T; (c) F; (d) T

REVIEW OF LOGIC

For frames 69-72, assume that p, q, and r are true propositions.

69. Write the truth value for each of the following disjunctions.

___(a) $\sim(p \vee \sim q)$
___(b) $p \vee q \vee \sim p \vee \sim q$

- - - - - - - - - - - - - -

(a) F; (b) T

70. Find the truth value for each of the following expressions.

___(a) $\sim p \wedge \sim q \wedge r$
___(b) $\sim(p \wedge q)$

- - - - - - - - - - - - - - -

(a) F; (b) F

71. Find the truth value of each of the following expressions.

___(a) ~p → (q ∧ ~r)
___(b) (~p ∨ r) → ~q

- - - - - - - - - - - - - - -

(a) T; (b) F

72. Find the truth value for each expression below.

___(a) (p ∧ ~q) ↔ (q ∧ r)
___(b) (r ∨ p) ↔ (q ∨ ~p ∨ ~r) ∧ r

- - - - - - - - - - - - - - -

(a) F; (b) T

73. Which of the following are called DeMorgan's rules?

___(a) ~(p ∨ q) ≡ ~p ∧ ~q
___(b) p ∨ (q ∨ r) ≡ (p ∨ q) ∨ r
___(c) p ∨ (q ∧ r) ≡ (p ∧ q) ∨ (p ∧ r)
___(d) ~(p ∧ q) ≡ ~p ∨ ~q

- - - - - - - - - - - - - - -

(a); (d)

74. Complete the truth table below to verify that
p ∧ (q ∨ r) ≡ (p ∧ q) ∨ (p ∧ r).

p	q	r	q ∨ r	p ∧ (q ∨ r)	p ∧ q	p ∧ r	(p ∧ q) ∨ (p ∧ r)
T	T	T					
T	T	F					
T	F	T					
T	F	F					
F	T	T					
F	T	F					
F	F	T					
F	F	F					

- - - - - - - - - - - - - -

p	q	r	q ∨ r	p ∧ (q ∨ r)	p ∧ q	p ∧ r	(p ∧ q) ∨ (p ∧ r)
T	T	T	T	T	T	T	T
T	T	F	T	T	T	F	T
T	F	T	T	T	F	T	T
T	F	F	F	F	F	F	F
F	T	T	T	F	F	F	F
F	T	F	T	F	F	F	F
F	F	T	T	F	F	F	F
F	F	F	F	F	F	F	F

$$\underbrace{\qquad\qquad} \equiv \underbrace{\qquad\qquad}$$

75. Write the converse and the contrapositive of the implication
    $(p \lor q) \to \sim r$.

    (a) converse _____

    (b) contrapositive _____

- - - - - - - - - - - - - -

(a) $\sim r \to (p \lor q)$; (b) $r \to \sim(p \lor q)$ or $\sim(\sim r) \to \sim(p \lor q)$

76. Which of these equivalences is valid?

    ___(a) $p \to q \equiv \sim q \to \sim p$
    ___(b) $p \to q \equiv q \to p$
    ___(c) $p \land q \equiv q \land p$
    ___(d) $p \lor q \equiv q \lor p$

- - - - - - - - - - - - - -

(a); (c); (d) (Choice (b) is a converse, which is not equivalent to its implication.)

Now take the Self-Test on the next page.

SELF-TEST

This Self-Test will help you evaluate whether or not you have mastered the chapter objectives and are ready to go on to the next chapter. Answer each question to the best of your ability. Correct answers are given at the end of the test.

1. Which of the following are propositions?

    ___(a) $5 + 9 = 15$
    ___(b) Fish are fascinating.
    ___(c) Are you happy?
    ___(d) All birds have feathers.

2. Give the truth value of each proposition below.

    ___(a) No fish can swim
    ___(b) $8^2 = 64$
    ___(c) All normal men have two legs.

3. Write a negation for each proposition in question 2.

    (a) _____

    (b) _____

    (c) _____

4. Write the symbol for each of the following.

    ___(a) negation
    ___(b) conjunction
    ___(c) disjunction
    ___(d) implication
    ___(e) strict implication
    ___(f) equivalence

5. Assume that p, q, and r are true propositions. Give the truth value of each expression below.

    ___(a) $(q \land \sim p) \lor (\sim r)$
    ___(b) $\sim(\sim p \lor r) \rightarrow (q \land p)$
    ___(c) $(\sim p \lor r) \leftrightarrow \sim(q \land \sim r)$
    ___(d) $(r \land p \land q) \rightarrow (\sim p \lor \sim r)$

6. Which of the equivalences below are called DeMorgan's rules of logic?

___(a) $p \leftrightarrow q \equiv q \leftrightarrow p$

___(b) $\sim(p \vee q) \equiv \sim p \wedge \sim q$

___(c) $p \vee (q \wedge r) \equiv (p \vee q) \wedge (p \vee r)$

___(d) $\sim(p \wedge q) \equiv \sim p \vee \sim q$

7. Construct a truth table to see if the equivalence below is valid.
$$p \wedge (q \vee r) \equiv (p \vee q) \wedge (p \vee r)$$

p	q	r	
T	T	T	
T	T	F	
T	F	T	
T	F	F	
F	T	T	
F	T	F	
F	F	T	
F	F	F	

8. Write a converse and a contrapositive for the implication $p \rightarrow (\sim q)$.

(a) converse _____

(b) contrapositive _____

ANSWERS TO SELF-TEST

Compare your answers to the Self-Test with the correct answers given below. If all of your answers are correct, you are ready to go on to the next chapter. If you missed any questions, study the frames indicated in parentheses following the answer.

1. (a); (d) (frames 1-3)

2. (a) F; (b) T; (c) T (frames 4-5)

3. (a) Some fish can swim. (frames 6-10)
    (b) $8^2 \neq 64$ (frames 6-10)
    (c) All normal men do not have two legs, or some normal man does not have two legs. (frames 6-10)

4. (a) $\sim$ (frame 8)   (d) $\rightarrow$ (frame 47)
    (b) $\wedge$ (frame 12)   (e) $\leftrightarrow$ (frame 62)
    (c) $\vee$ (frame 20)   (f) $\equiv$ (frame 34)

5. (a) F (frames 8, 13, and 20)   (c) T (frames 62 and 68)
    (b) T (frames 47-49)   (d) F (frames 47-49)

6. (b); (d) (frames 33-44)

7.

p	q	r	q $\vee$ r	p $\wedge$ (q $\vee$ r)	p $\vee$ q	p $\vee$ r	(p $\vee$ q) $\wedge$ (p $\vee$ r)
T	T	T	T	T	T	T	T
T	T	F	T	T	T	T	T
T	F	T	T	T	T	T	T
T	F	F	F	F	T	T	T
F	T	T	T	F	T	T	T
F	T	F	T	F	T	F	F
F	F	T	T	F	F	T	F
F	F	F	F	F	F	F	F

p $\wedge$ (q $\vee$ r) $\not\equiv$ (p $\vee$ q) $\wedge$ (p $\vee$ r)   The equivalence is not valid.
(frames 41-42)

8. (a) $(\sim q) \rightarrow p$ (frames 57-59)

   (b) $q \rightarrow (\sim p)$ or $\sim(\sim q) \rightarrow (\sim p)$ (frames 53-56)

CHAPTER FOUR
Algorithms and Flowcharts

In this chapter, you will learn to use the flowchart, one of the basic tools of computer work. The flowchart traces a path that you, as a programmer, will instruct the computer to follow. This text is not concerned with teaching you how to draw a flowchart, although you may be able to do so. We will use flowcharts extensively in later chapters and you will learn to read them here. By the end of this chapter you will be familiar with flowchart conventions and language. You will be able to focus later on the functions of flowcharts, and not be concerned with their mechanics.

When you complete this chapter, you will be able to:

- name the necessary features of a flowchart or algorithm;

- name some desirable features of algorithms or flowcharts;

- determine if a given flowchart has these features;

- identify standard flowchart symbols;

- interpret logical expressions in flowcharts;

- trace a path of given data through a flowchart;

- determine if flowcharts or algorithms are equivalent.

If you have studied or worked with flowcharting before, you may feel you already meet these objectives. In that case, turn directly to frame 53, where the review exercises begin.

ALGORITHMS

1. Computers must be given detailed instructions before they can effectively help you solve your problems. The detailed instructions make up an <u>algorithm</u>, a step-by-step procedure that the computer can carry out. Which of the following statements is true?

    ___(a) An algorithm cannot be carried out by a computer.
    ___(b) An algorithm can solve a problem.
    ___(c) An algorithm gives instructions one at a time for solving a problem.

    - - - - - - - - - - - - - - -

    (c)

2. Below is a sample algorithm which could be used to find the fourth power of a number called x.

    1. Give a specific value to x.
    2. Multiply x times x.
    3. Multiply that result times x.
    4. Multiply that result times x.
    5. Write that result as the value of x^4.

    (a) The sample algorithm above is a _____-____-_____ procedure.

    (b) The algorithm tells you what instructions to give to the

    _____.

    - - - - - - - - - - - - - -

    (a) step-by-step; (b) computer

3. The algorithm in frame 2 is not very efficient, even though it is effective in finding the value of x^4. Which of the algorithms below would find and write the answer?

    ___(a) 1. Give a value to x.
         2. $y = x^4$
         3. Write the value of y.

    ___(b) 1. Give a value to x.
         2. Multiply x times 4.
         3. Write the result.

    ___(c) 1. Give a value to x.
         2. Multiply x times x times x times x.
         3. Write the result.

- - - - - - - - - - - - - -

(a); (c) (Most computer languages supply algorithms which find powers of a number, so (c) would not be used.)

BASIC FLOWCHARTING SYMBOLS

4. Look at the algorithm and corresponding flowchart symbols below. The flowchart gives the same instructions, but in diagram form.

Algorithm

1. Give a value to x.
2. $y = x^4$
3. Write the value of y.

Flowchart

(a) Which instruction in the algorithm is represented in the flowchart as a computer (IBM)
card? _____

(b) Which instruction in the algorithm is represented in the flowchart as a symbolic printer
page? _____

(c) What shape symbol is used for processing steps such as algebraic equalities? _____

- - - - - - - - - - - - - -

(a) 1. Give a value to x.; (b) 3. Write the value of y.;
(c) a rectangle

5. In computer programming, values are frequently given to variables
    by input cards. The value of a variable changes, or varies, for
    each problem. The variable x may be used for a salary in one pro-
    blem, for a number to be cubed in another. In flowcharts, the sym-
    bol for card input is a card outline. Inside the card outline we write
    the names of variables whose values will be on the actual card used
    as input. Which of the following might be an acceptable input sym-
    bol for a flowchart?

___(a) $y = x^4$ ___(b) AMOUNT   ___(c) AMOUNT

- - - - - - - - - - - -

(b)

6. Flowcharts are diagrammed algorithms. If the algorithm has an
    instruction like "give a value to the variable," the flowchart might
    indicate input of the value with a symbol shaped like _____.
    Written inside the symbol would be the name of the _____.

- - - - - - - - - - - -

a card (or ⬡); variable

7. Algorithms usually include processing steps. These are instruc-
    tions such as "$y = x^4$," or "find 5% of the total," or "calculate the
    result of A divided by B." Processing steps in flowcharts are en-
    closed in rectangles, with the processing instructions inside the
    rectangle. Which of the following might represent a processing
    step in a flowchart?

___(a) x   ___(b) $y = x^4$ ___(c) $y = x^4$

- - - - - - - - - - - -

(c)

8. Output information from the computer tells you what it found out. In
    our sample algorithm, it would do no good for the computer to cal-
    culate $y = x^4$ if it did not let you know what the answer was. Of
    course, computers have many ways of communicating with program-
    mers. A display tube can show answers. The typewriter in the
    computer room can also produce output. Computers can punch
    cards to give results, or produce output directly on a teletype ma-
    chine. But you will probably want your output on a printed page.
    For this reason we will use the outline of the printer page as an out-
    put symbol throughout this text. The name of the variable whose
    value is to be printed is written inside the output symbol.

    Which of the following might be used as an output portion of a
    flowchart?

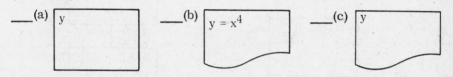

(c) (The standard programming symbol for all input and output is
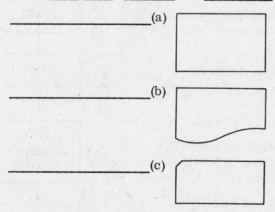
. We are using system symbols to make it easy for
you to tell what is input and what is output.)

9. Write <u>input card</u>, <u>printout</u>, or <u>processing</u> before each symbol below.

    _____(a)

    _____(b)

    _____(c)

- - - - - - - - - - - - - -

(a) processing; (b) printout; (c) input card

10.   This symbol, the oval, is always used at the beginning and end of a complete flowchart. In it you
write START or STOP, depending on where the oval is. When you
write a program working from a flowchart, the oval reminds you to
write the instructions to start and stop the computer's actions.

Which flowchart below is complete?

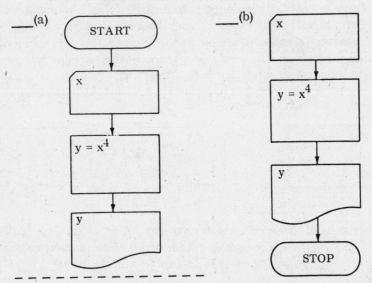

___(a)   START   ___(b)   x

             x $y = x^4$

             $y = x^4$ y

             y   STOP

neither (Ovals for both START and STOP must be included. These
are called flowchart segments, since they are not complete.)

11. The flowchart on the right represents an algorithm. Which of the algorithms below does it
represent?

___(a) 1. Give a value to TOTAL.
        2. Find values of ONHAND and
            INCOME.
        3. Print value of INCOME.

___(b) 1. Print ONHAND and INCOME.
        2. Add ONHAND to INCOME.
        3. Give a value to TOTAL.

___(c) 1. Give values to ONHAND and
            INCOME.
        2. Add ONHAND and INCOME.
        3. Print value of TOTAL.

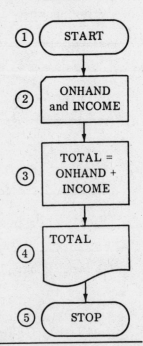

① START

② ONHAND and INCOME

③ TOTAL = ONHAND + INCOME

④ TOTAL

⑤ STOP

– – – – – – – – – – – – – –

(c)

12. Write the number of the flowchart symbol in frame 11 that is used for each of the following reasons.

___(a) Stop the computer
___(b) Start the computer
___(c) Output information
___(d) Input information
___(e) Processing steps

– – – – – – – – – – – – – –

(a) 5; (b) 1; (c) 4; (d) 2; (e) 3

13. Match the flowchart symbols below with the correct meanings.

___(a)

1. Processing steps

2. Input cards

___(b)

3. Printout

4. Start

5. Stop

___(c)

___(d)

– – – – – – – – – – – – –

(a) 4, 5; (b) 2; (c) 3; (d) 1

FLOWCHARTING OF DECISIONS

14. In a program, we ask computers to make decisions. The processing or output will vary depending on conditions. An algorithm and
corresponding flowchart are illustrated below.

Algorithm   Flowchart

1. Get the value of variables
   A and B.
2. Compare the two variables.
3. If A is larger, print A.
4. Otherwise, print B.

Study the flowchart carefully.
Which variable will be printed
in each of the following cases?

___(a) A = 3, B = 4

___(b) A = 4, B = 3

___(c) A = 4, B = 4

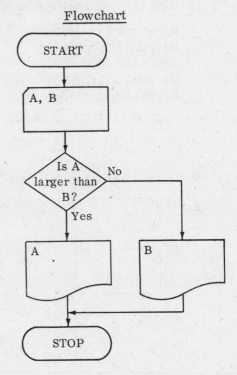

- - - - - - - - - - - - - -

(a) B; (b) A; (c) B (A is not larger than B, so B will be printed.)

15. Look again at the flowchart in frame 14. What shape is used in

flowcharts to represent decision points? _____

- - - - - - - - - - - - - -

a diamond shape

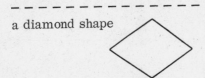

16. The question "is A larger than B" can be asked mathematically in
several ways. "A >B" is one way that uses a proposition. Once
you give values to A and B, you can easily determine the truth value
of this proposition. Which flowchart segment below represents this
method of comparing A and B for the algorithm of frame 14?

___(a)

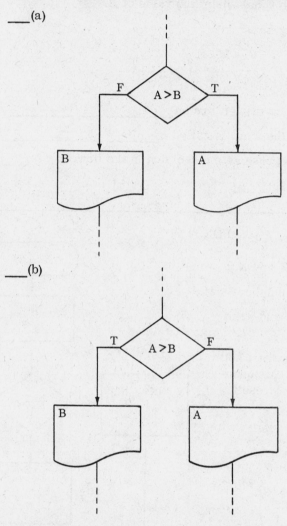

___(b)

- - - - - - - - - - - - -

(a) (If you ask a question in your decision box, as in frame 14, the
alternative branches are called Yes and No. If you use a proposition, as here, the branches are labeled with truth values.)

17. Which statement below explains the flowchart segment (b) in frame 16?

___(a) Print the value of A or B, whichever is larger.
___(b) Print the smaller of the two values A and B.
___(c) If A and B are equal, print the value of B.

- - - - - - - - - - - - - -

(b)

18. Look at Box A.

(a) The steps listed on the left are an _____.

(b) The diagram on the right is called a _____.

(c) Which number represents the decision in the flowchart? _____

in the algorithm? _____

BOX A

1. Give values to N1 and N2.

2. Is N1 a positive number?

3. If so, square N1 and print the result.

4. If N1 is 0 or negative, add it to N2, square this quantity, and print the result.

① START

② N1, N2

③ N1>0

(4a) $N = (N1 + N2)^2$ F

T

(4b) $N = N1^2$

⑤ N

⑥ STOP

- - - - - - - - - - - - - -

(a) algorithm; (b) flowchart; (c) 3 ($N1 > 0$), 2 (Is N1 a positive number?)

19. Refer again to Box A and give the value that will be printed in each of the following cases.

___(a) N1 = -1, N2 = 2
___(b) N1 = 2, N2 = -1
___(c) N1 = 0, N2 = -2

- - - - - - - - - - - - -

(a) 1 $[(-1 + 2)^2]$; (b) 4 $[(2)^2]$; (c) 4 $[(0 + (-2)^2]$

20. You have seen some examples of branching depending on a decision; the program branches to one point or another depending on the outcome of the decision. Decisions and branches can become more complex, and may require several decisions.

For example, suppose we wish the computer to make a decision based on the value of a logical conjunction, $A > 0 \wedge B > 0$. If the conjunction is true, the computer is to print the value of $(A + B)^2$, called C. If the conjunction is false, the computer will just print the values of A and B. We will diagram this using two decision blocks.

Turn the page and use the algorithm on the left to complete the flowchart on the right.

Algorithm	Flowchart

1. Give values to A and B.

2. Is A > 0?

3. If so, see if B > 0.

4. If A ≯ 0, print values of A and B.

5. Is B > 0?

6. If so, print $(A + B)^2$ (call it C).

7. If B ≯ 0, print values of A and B.

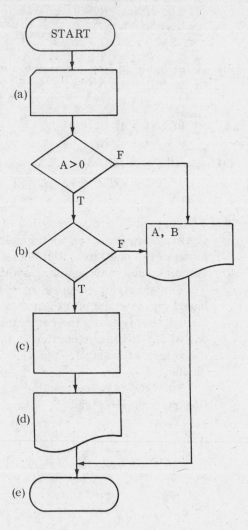

- - - - - - - - - - - - - -

(a) A, B; (b) B > 0; (c) $C = (A + B)^2$; (d) C; (e) STOP

21. The flowchart in frame 20 could also be drawn using only one decision block, as shown below. What would you write in the decision block here? _____

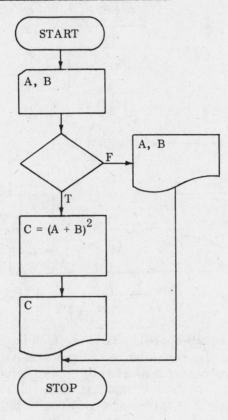

- - - - - - - - - - - - - - -
A > 0 AND B > 0, or A > 0 ∧ B > 0 (or an equivalent expression)

22. The flowchart below represents a situation in which the computer
will read one card as input, make a decision, then print out the
higher of the two values and stop.

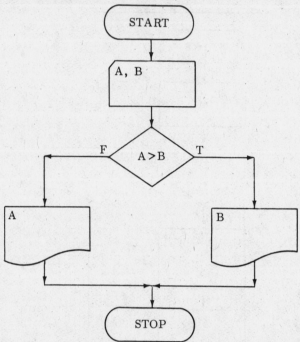

A more likely situation would be one in which the programmer has a
whole pile of cards, each with two numbers on it. He would want to
have the computer print the higher of the two values from each card
before stopping. To do this, he must make his flowchart loop back
to an earlier point. He also must know how to tell the computer to
stop reading cards.

A stack of cards with the same layout can be called a card file.
At the end of the stack, the programmer puts a special card called
an end-of-file card (EOF). This card tells the computer that it has
reached the end of that card file. The programmer can test for this
card, and branch to another action or to the end of his program, depending on circumstances.

The flowcharts on the next page in Box B are labelled I and II.
Study them carefully.

(a) Which flowchart represents an algorithm for handling only one

   card? _____

(b) Which flowchart includes a decision about an end-of-file card?

   _____

(c) Which flowchart will handle three cards or 300 cards? _____

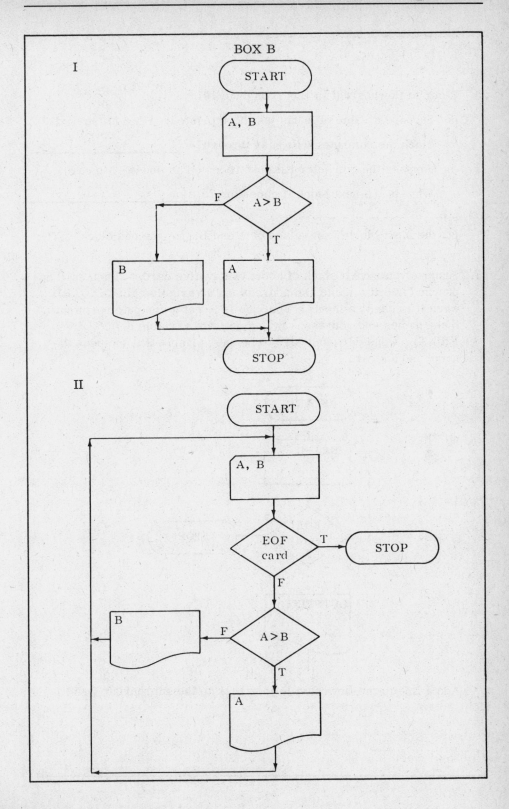

_ _ _ _ _ _ _ _ _ _ _ _ _ _

(a) I; (b) II; (c) II

23. Refer to flowchart II in Box B on page 101.

(a) Suppose a value of A has just been printed. What is the next block the computer will pass through? _____

(b) Suppose the computer has just received an end-of-file card. What is the next block it considers? _____

_ _ _ _ _ _ _ _ _ _ _ _ _ _

(a) the input block: get values of A and B; (b) the end block: STOP

24. Suppose you have a stack of about twenty-five cards. Each card has punched into it a name and address (as a variable called INCARD when it is used as input). Your job is to get a computer printout of these names and addresses (as an outprint variable called OUT-PRINT). Consider each card to be one variable that includes the name and address.

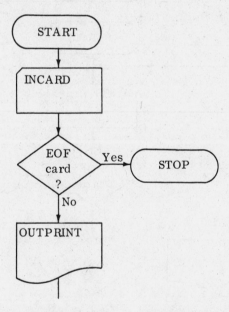

Add a line to the flowchart to loop back to the appropriate point.

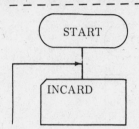

(The flowchart must loop back to between the START block and the input block. It may be on either side of the flowchart.)

COUNTERS IN FLOWCHARTS

25. A statement like "x = x + 1" is a reasonable statement to a computer. It may also be written as "add 1 to x." In either case, the computer replaces its current value for x with the value of x + 1. This fact is frequently used by programmers in building counters into programs.

  Suppose your flowchart indicates that x = x + 1 immediately after each input card is received. What does the statement do to the input cards?

___(a) causes them to be received
___(b) changes the value on the cards
___(c) counts the input cards as they are received

- - - - - - - - - - - - - - -

(c)

26. The flowchart below is much like the one you completed in frame 24.

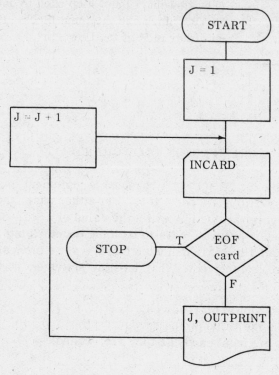

Which of the statements below explains the difference in this flow-chart?

___(a) This chart counts the input cards.

___(b) This chart does not cause the names and addresses to be printed.

___(c) The output information is numbered as it is printed.

- - - - - - - - - - - - - - -

(a); (c) (The value of the counter J is set to 1 for the first card. Then it is increased by 1 for each additional card.)

27. Suppose you have a problem similar to the one charted in frame 26. This time, however, you want to number and print only the first twenty names. You want the computer to stop acting either after it prints twenty names, or when it reaches the end-of-file card.

(a) Fill in the blank decision box in the flowchart below.
(b) Draw a line to show where the path goes after J = J + 1.

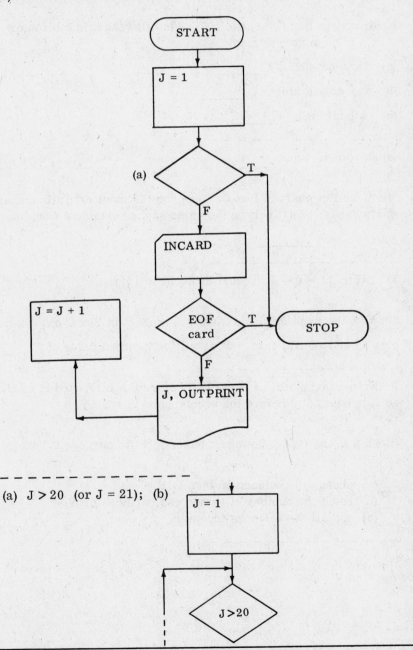

– – – – – – – – – – –

(a) J > 20 (or J = 21); (b)

28. In the flowchart in frame 27, the variable J was used as a:

___(a) signal for the end-of-file card
___(b) counter

- - - - - - - - - - - - - - -

(b)

29. Refer now to Box C on page 107. This flowchart is a bit more complex. It contains how many:

(a) decision blocks? _____

(b) processing blocks? _____

(c) output blocks? _____   ·

- - - - - - - - - - - - - -

(a) 2; (b) 4; (c) 2

30. Study the flowchart in Box C. Suppose the input card file contains thirty cards. What will be the value of the variable J when the computer stops? _____

- - - - - - - - - - - - - -

11 (When $11 - J = 0$, J must be equal to 11.)

31. Suppose the input card file for the flowchart in Box C has six cards plus an end-of-file card. What will be the final value of J? _____

- - - - - - - - - - - - - -

7 (Notice that J will be equal to 7 before the EOF card is read, just as it is equal to 2 before the second card is read.)

32. Block 6 in the Box C flowchart has a specific purpose. That purpose is to:

___(a) change the balance on that card
___(b) count the number of names and balances printed
___(c) add all the balances together

- - - - - - - - - - - - - -

(c)

BOX C

Variables

J = counter
TBAL = total balance
NAME = customer's name

BALANCE = customer's individual
balance
AVERAGE = total of all customer's
balances divided by the number
of customers

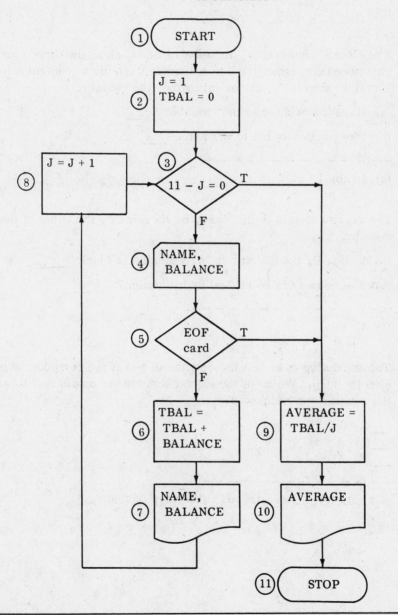

33. Which of the descriptions below describes best the output you might expect from a program based on the chart in Box C?

___(a) Ten names will be listed, with the average balance for each.
___(b) First up to ten names will be listed, each with a balance. Then the average of the ten balances will be printed.
___(c) The values of J from 1 to 10 and TBAL from 0 to its highest value will be printed.

- - - - - - - - - - - - - -

(b)

34. The blocks numbered 2, 3, and 8 in Box C show the three components necessary when you use a counter variable to control a loop. The first step is to set, or initialize, the counter.

(a) Here the counter is the variable _____.

(b) The counter is set to the value _____.

- - - - - - - - - - - - - -

(a) J; (b) 1

35. The second step is to increase, or increment, the value of the counter.

(a) In Box C, the counter is incremented in block _____.

(b) The counter is increased by how much? _____

- - - - - - - - - - - - - -

(a) 8; (b) 1

36. The third step is to test the counter to see if the computer should stop the loop. Which of the expressions below would test to see if the counter has reached 31?

___(a) J > 30
___(b) J = 31
___(c) 31 − J = 0

- - - - - - - - - - - - - -

(a); (b); (c) (These are all valid ways of testing.)

37. The steps in using a counter to control a loop are listed below.
    Number them in the correct order, using 1 for the first step.

    ___(a) Test the counter
    ___(b) Initialize, or set, the counter
    ___(c) Increment, or raise, the counter

    _ _ _ _ _ _ _ _ _ _ _ _ _ _

    (a) 3; (b) 1; (c) 2

38. Match the steps with the examples.

    ___(a) Test the counter   1. J = 1
    ___(b) Initialize the counter   2. J > 40
    ___(c) Increment the counter   3. J = J + 2

    _ _ _ _ _ _ _ _ _ _ _ _ _ _

    (a) 2; (b) 1; (c) 3

39. List in order the three steps in using a counter to control a loop.

    (a) _____

    (b) _____

    (c) _____

    _ _ _ _ _ _ _ _ _ _ _ _ _ _ _

    (a) Initialize (or set) the counter; (b) Increment (or raise) the
    counter; (c) Test the counter

You have learned a good deal about flowcharts already. You know that a
flowchart is a diagram of an algorithm, or a detailed way to solve a problem. You know the standard symbols used in flowcharting, how counters
can be used to control loops, and how branching is used to determine what
step is performed next depending on the outcome of a decision.

    In an important sense, a flowchart is an algorithm. In the next sec-
tion we will talk about features of algorithms, but keep in mind that all
of these features also apply to flowcharts.

FEATURES OF ALGORITHMS

40. It is absolutely essential that algorithms be <u>unambiguous</u>. They must be clear; there must be no doubt at all about how to do a step or what step comes next. A step such as "multiply x by a smaller number" is ambiguous. A better way of writing such a step would be:

___(a) multiply x by a larger number
___(b) multiply x by the next smaller whole number
___(c) multiply x by one-half the value of x

- - - - - - - - - - - - - - -

(b); (c) (Neither of these is ambiguous if x has a value.)

41. An algorithm must also be <u>effective</u>. It must solve the problem. If the problem is to calculate net pay (take home), given gross pay (total earned) and assorted deductions, then the algorithm must handle all the facts and do things in the correct order to be effective. Which of the brief algorithms below might effectively solve this problem?

___(a) 1. Get values of gross pay and deductions.
       2. Add deductions together.
       3. Print result.

___(b) 1. Get values of gross pay and deductions.
       2. Add deductions to gross pay.
       3. Print result.

___(c) 1. Get values of gross pay and deductions.
       2. Add deductions together.
       3. Subtract total of deductions from gross pay.
       4. Print result.

- - - - - - - - - - - - - - -

(c)

42. An algorithm must be clear and definite. There must be no doubt about what is to be done when. This means that an algorithm must

be _____.

- - - - - - - - - - - - - - -

unambiguous

43. An algorithm must solve the problem it is designed for. This means that an algorithm must be _____.

- - - - - - - - - - - - - - - -

effective

44. It is desirable that an algorithm be general and efficient. An algorithm that is not general or very efficient can still be acceptable if it has two required features. Write the necessary features.

_____   _____

- - - - - - - - - - - - - - - -

unambiguous; effective (in either order)

45. We say that an algorithm should be general. This means that an algorithm should be able to cover all possibilities. It should not be too specific or limited in the problems it solves. Which of these descriptions of an algorithm is most general?

___(a) The algorithm can handle averaging of up to 100 test scores.
___(b) The algorithm can only handle the averaging of exactly 100 test scores.
___(c) The algorithm can handle the averaging of any number of test scores.

- - - - - - - - - - - - - - -

(c)

46. Which of the three descriptions of algorithms in frame 45 is least general? _____

- - - - - - - - - - - - - - -

(b) (This is the most specific, therefore the least general.)

47. Another desirable feature of an algorithm is that it be efficient. Suppose you have two algorithms for multiplication. One uses a multiplication table; the other uses repeated addition. Which is more efficient? _____

- - - - - - - - - - - - - - -

the one that uses a multiplication table

48. Which of the following are not necessary in an algorithm, but are desirable?

___(a) effectiveness
___(b) unambiguity
___(c) efficiency
___(d) generality

- - - - - - - - - - - - - - - -

(c); (d)

49. In frame 47, we mentioned two algorithms to solve the same problem, but one was more efficient than the other. These two algorithms are equivalent because they solve the same problem, or have the same function.

    Suppose you have two algorithms for finding square roots. The first will find the square root of any whole number. The second will find the square root of any whole number smaller than 20,000.

(a) Which example is more general, the first or the second?

    _____

(b) Suppose your task is to find square roots of all whole numbers between 100 and 1000. Are the two examples equivalent?

    _____

- - - - - - - - - - - - - - - -

(a) first (It isn't limited in the size of the number.); (b) yes (They solve the same problem, that of finding square roots. They would not be equivalent if you needed to find the square root of a whole number greater than 20,000.)

50. List the two required features for an algorithm.

    _____   _____

- - - - - - - - - - - - - - - -

unambiguous; effective (in either order)

51. List two additional desirable features for an algorithm.

    _____   _____

- - - - - - - - - - - - - - - -

general; efficient (in either order)

52. Two algorithms may be said to be equivalent if they:

___(a) have the same level of generality
___(b) are equally efficient
___(c) effectively solve the same problem

— — — — — — — — — — — — — —

(c)

FLOWCHART REVIEW AND EXPANSION

For the remainder of this chapter, we shall review and expand what you
have already learned. You will be given fairly complex flowcharts and
asked questions about various parts of them. If you can answer these
questions, you will have acquired the ability to read and interpret flowcharts, at least to the extent that will be required in the rest of this
book. For this reason, no separate Self-Test is included for this chapter. Now begin the review exercises with frame 53 and Box D.

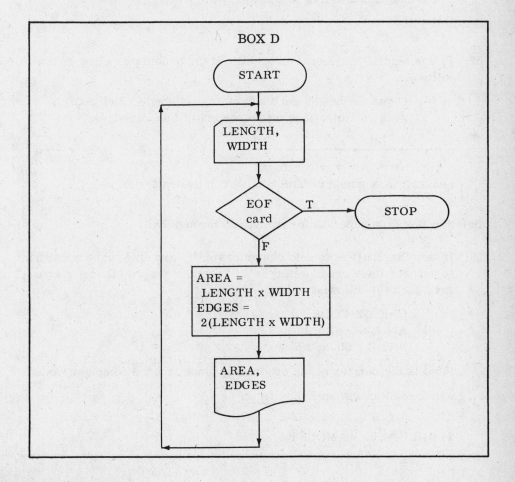

Refer to Box D on page 113 for frames 53 through 55.

53. The algorithm represented by this flowchart solves the problem of:

___(a) finding and printing the length and width of rectangles
___(b) finding and printing the area and edge measurement of rect-
angles
___(c) calculating the largest sized rectangle

— — — — — — — — — — — — — — —

(b)

54. Which of the following are included in Box D?

___(a) a decision block
___(b) a loop
___(c) a branch
___(d) a counter

— — — — — — — — — — — — — — —

(a); (b); (c)

55. Is this algorithm general or specific if the problem is given as
follows:

Given the length and width of any rectangle, find the
area and total edge measurement of that rectangle.

_____

— — — — — — — — — — — — —

general (Any number of rectangles can be dealt with.)

Refer to Box E on page 115 for frames 56 through 58.

56. In this flowchart segment, customers' bills are discounted according
to whether their credit rating is 1, 2, or 3. Suppose the first input
card gave the following values:

CREDIT-CODE = 1
AMOUNT = 20.00
NAME = MORGAN

What is the number of the processing block that the computer would

use for this customer? _____

— — — — — — — — — — — — — — —

1 (BILL = .90 x AMOUNT)

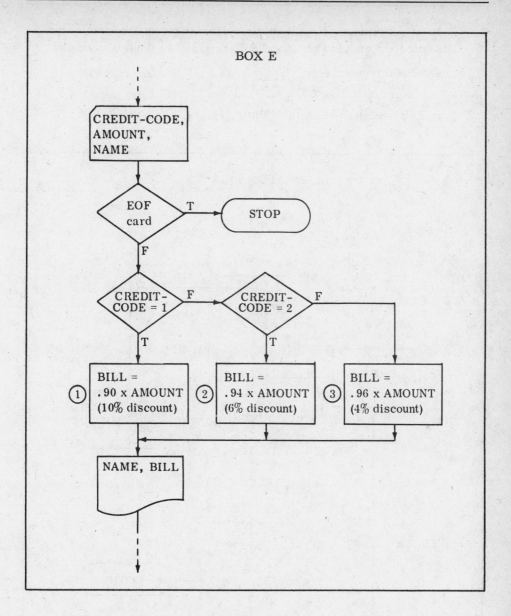

BOX E

57. What would be printed for the customer in frame 56?

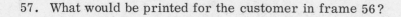

MORGAN   18.00

58. Suppose a keypuncher made an error in preparing the input cards. She entered 7 instead of 1 for CREDIT-CODE. What processing block would be used now? _____

- - - - - - - - - - - - - - -

3 (CREDIT-CODE is neither 1 nor 2.)

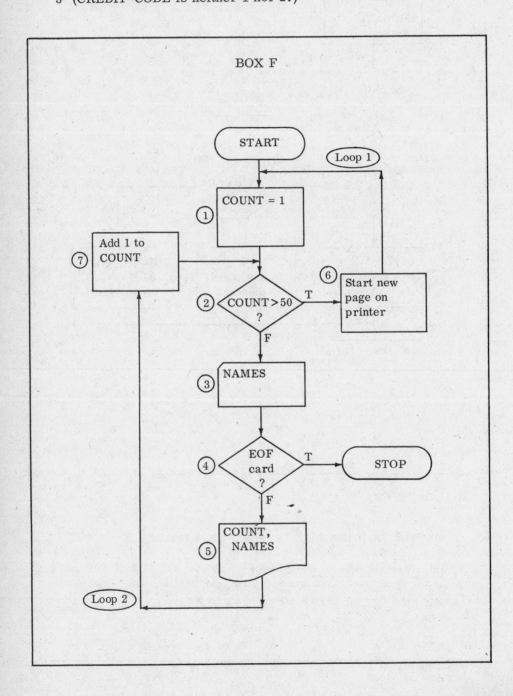

BOX F

Refer to Box F on page 116 for frames 59 through 61.

59. The flowchart in Box F uses how many loops? _____

_ _ _ _ _ _ _ _ _ _ _ _ _ _ _ _

2

60. The purpose of Loop 1 is to:

___(a) start reading input cards from the beginning again
___(b) start a new printer page and renumber from 1
___(c) increment the counter

_ _ _ _ _ _ _ _ _ _ _ _ _ _ _ _

(b)

61. Most of the blocks in the flowchart in Box F are numbered. Give the function of each block listed below.

(a) Block 1 _____

(b) Block 2 _____

(c) Block 4 _____

(d) Block 7 _____

_ _ _ _ _ _ _ _ _ _ _ _ _ _ _ _

(a) set or initialize the counter (processing)
(b) test the counter (decision)
(c) test for end-of-file card (decision)
(d) raise or increment the counter (processing)

Refer to Box G on page 118 for frames 62 through 66.

62. The flowchart in Box G contains a nested loop (a loop entirely within another loop). Study this flowchart.

Suppose that AGE can be any whole number from 12 to 99. SEX is 1 for females and 2 for males. This problem results in a printout of:

___(a) all males over 18
___(b) all males under 18
___(c) all females over 18
___(d) all females under 18

_ _ _ _ _ _ _ _ _ _ _ _ _ _ _ _

(c)

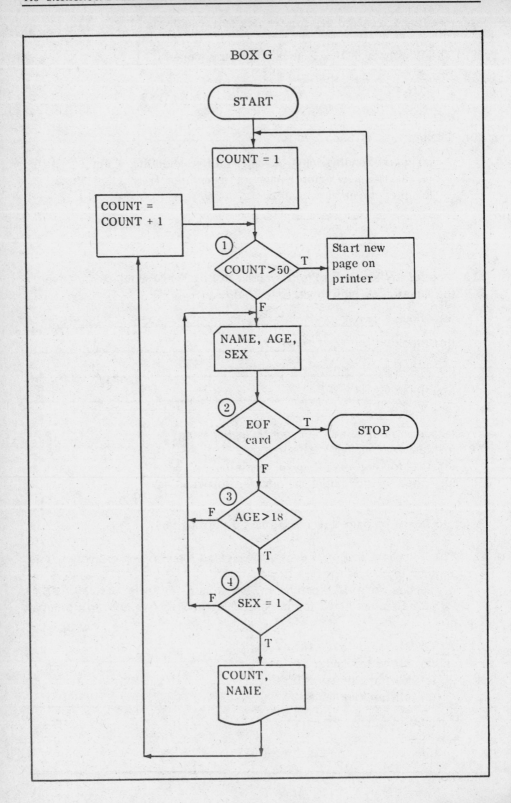

63. The counter in this flowchart is incremented:

   ___(a) immediately before a card is read
   ___(b) immediately after a line is printed
   ___(c) each time a female over 18 is located

(b); (c)

64. Blocks 3 and 4 could be replaced by a single decision block. What
would you write in that single block? _____

AGE $> 18 \wedge$ SEX $= 1$, or AGE > 18 AND SEX $= 1$

65. (a) The flowchart in Box G can handle how many input cards?

   _____

(b) One printer line is printed each time the output block is performed. How many lines can be printed, based on this flowchart? _____

(a) any number; (b) any number, but only fifty per page

66. Compare the flowcharts in Box F and Box G. Are they equivalent?

   _____

----

no (The flowchart in Box G solves a different problem.)

Refer to Box H on page 120 for frames 67 through 71.

67. Study the flowchart in Box H. Suppose the input card sets A = 3,
B = 4, and C = 6. Will the resulting truth value of block 1 be T or

F? _____

F

68. With the same input information as in frame 67, the path then goes
to block 2. From there, the next block on the path is numbered

   _____.

_ _ _ _ _ _ _ _ _ _ _ _ _

5

69.   For this information, what is the value of ANSWER? _____

_ _ _ _ _ _ _ _ _ _ _ _ _ _

6 (or C)

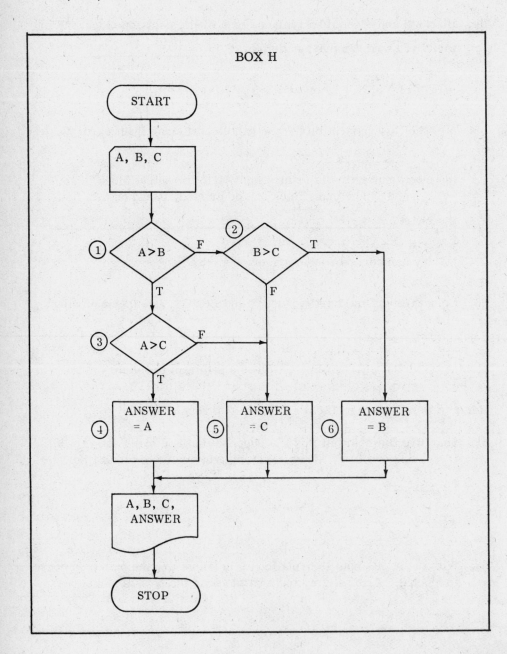

70. The purpose of this flowchart is to find:

___(a) the sum of three numbers
___(b) the smallest of three numbers
___(c) the greatest of three numbers

– – – – – – – – – – – – – – –

(c)

71. How many sets of three numbers can this flowchart accept before

stopping? _____   Is the flowchart general or specific?

_____

– – – – – – – – – – – – – –

one; specific

Refer to Box I on page 122 for frames 72 through 75.

72. Study the flowchart in Box I.

(a) Does it seem unambiguous? _____

(b) Does it seem effective? _____

(c) Does it seem general? _____

(d) Does it seem reasonably efficient? _____

– – – – – – – – – – – – – –

(a) yes; (b) yes; (c) yes; (d) yes
(These are relative answers. It is entirely possible that other algorithms might be more efficient, effective, general, or unambiguous.)

73. List the block numbers between 1 and 6 on the path of this set of

variable values: A = 7, B = 8, C = 6. _____

– – – – – – – – – – – – – –

1, 2, 6

74. List the blocks on the path of A = 8, B = 7, C = 9. _____

– – – – – – – – – – – – – –

1, 3, 5

75. Compare the flowchart in Box I with the one in Box H.   Are they equivalent? _____

   _ _ _ _ _ _ _ _ _ _ _ _ _ _

   yes

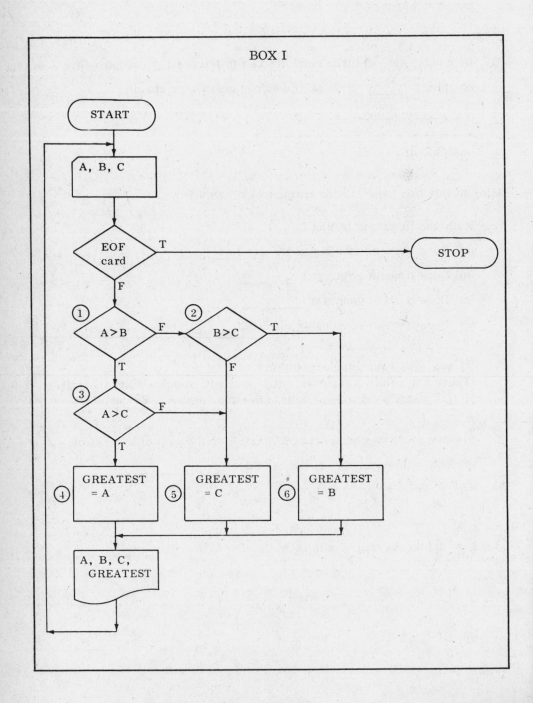

BOX I

CHAPTER FIVE

Computer Arithmetic and Notation

Computers use numbers in different forms. Keeping track of decimal points is a problem for computers just as it is for people. As a programmer, you must tell the computer if and where each input and output constant (specific value) or variable has a decimal point. In this chapter, we will study whole numbers, and what the computer can do with them, and we will look at floating-point numbers with their decimal points. We will then learn a new way of writing decimal fractions in E notation and how to do arithmetic with numbers in this form. An understanding of the advantages and defects of these techniques will help you in later math and computer work.

When you complete this chapter, you will be able to:

- identify integers;

- add, subtract, and multiply integers;

- divide integers, and find the remainder;

- identify floating-point and fixed-point numbers;

- write numbers in E notation;

- normalize numbers in E notation;

- add and subtract numbers written in E notation;

- multiply and divide numbers written in E notation;

- identify the effect of chopping or rounding by the computer;

- identify mantissas and exponents in E notation.

INTEGER ARITHMETIC

1. Integers are whole numbers with no fractional parts. The numbers $6\frac{1}{2}$, 0.4, and 6.0 are not integers. Which of the following are integers?

___(a) 5.2   ___(e) –3.2
___(b) $5\frac{1}{4}$ ___(f) –108
___(c) 67   ___(g) 0
___(d) 3.14159   ___(h) $-8\frac{1}{2}$

- - - - - - - - - - - - - - -

(c); (f); (g)

2. You use integer arithmetic in everyday life. Addition, subtraction, and multiplication of integers by a computer has the same results as when you do it. Give the answers to the integer arithmetic problems below.

   (a) 5 + 7 = _____

   (b) 17 – 9 = _____

   (c) 17 x (–8) = _____

- - - - - - - - - - - - - - -

(a) 12; (b) 8; (c) –136

3. In division, integer arithmetic varies from what you are used to. Since fractions are not used, any remainder is dropped. For example, 8 divided by 3 would be 2 in integer arithmetic. Give the results of the integer divisions below.

   (a) 5/2 = _____

   (b) 2/5 = _____

   (c) 100/17 = _____

- - - - - - - - - - - - - - -

(a) 2; (b) 0; (c) 5

4. I/J means I divided by J. I and J are integer variables—they can be replaced by any integers. We can use the expression I/J to represent an integer division. If I = 7 and J = 4, then I/J = _____.

- - - - - - - - - - - - - - -

1 (I/J = 7/4 = 1)

5. In integer division any remainder is lost—but you can find it again. The formula I − [(I/J) x J] = R means Numerator − (Quotient x Denominator) = Remainder. Use the formula to find the remainder lost in the problem in frame 4.

I − [(I/J) x J] = R

_____ − [(____/____) x ____] = _____

− − − − − − − − − − − − − − −

7 − [(7/4) x 4] = 7 − 4 = 3 (Of course, you can tell by just looking that 3 is the remainder, but the computer has to carefully calculate these things.)

6. When you perform integer division by hand, you do not need a formula to find the remainder. The computer, however, divides in a different way, and most likely in a different number system. Suppose now that I = 121 and J = 12. Find the value of I/J.

_____

− − − − − − − − − − − − − −

10

7. Use the formula I − [(I/J) x J] = R to find the remainder of the remainder of the integer division in frame 6.

_____ = _____

− − − − − − − − − − − − − −

121 − (10 x 12) = 1

8. Integer arithmetic differs from standard arithmetic in the operation of _____.

− − − − − − − − − − − − − −

division

9. In integer arithmetic, as in standard arithmetic, division by zero is not permitted. Actual division by numbers getting closer to zero results in an infinitely (uncountably) large number which no computer can handle. Computers usually stop operation when they try to divide by zero.

    Give the result of each integer arithmetic problem below, if there is one.

    (a) 0 + 6 = _____   (d) 0/2 = _____

    (b) 0 − 7 = _____   (e) 2/0 = _____

    (c) 0 x 17 = _____

    − − − − − − − − − − − − − −

    (a) 6; (b) -7; (c) 0; (d) 0; (e) An infinitely large number cannot be obtained, so the computer stops.

10. When is division by zero permitted?

    ___(a) in standard arithmetic
    ___(b) in integer arithmetic
    ___(c) never

    − − − − − − − − − − − − − −

    (c)

FLOATING-POINT ARITHMETIC

11. In computer terminology, integers are called <u>fixed-point</u> numbers. The imaginary point is fixed after the rightmost digit. Numbers that include a decimal point are called <u>floating-point</u> numbers, no matter where the point is placed. Label the numbers below as fixed-point or floating-point.

    (a) 2.7 _____

    (b) 27 _____

    (c) 27. _____

    (d) 0.27 _____

    − − − − − − − − − − − − − −

    (a) floating-point; (b) fixed-point; (c) floating-point (This contains a decimal point and is a floating-point number, even though it is a whole number.); (d) floating-point

When you program a computer, you determine whether a number (variable or constant) is fixed-point or floating-point. You also determine how many digit positions the number can have. How you do this depends on the computer and the language you are using. Fixed-point arithmetic is more efficient for the computer, but floating-point arithmetic can handle fractions and division much more effectively. Most of the arithmetic you ask a computer to do will be floating-point, as when you are working with prices, percents, or proportions. Sometimes, as when you use a counter, you will use integer arithmetic. If you then wish to divide by the value of the counter, you can easily convert it to a floating-point format. The method, of course, depends on the machine and programming language.

12. You determine the size of variables used in your program. Suppose your program is concerned with a payroll problem. Since everything is concerned with dollars and cents in the program, you may decide to allow two spaces to the right of the decimal point in hours and rate. You multiply HOURS x RATE to get GROSS.

(a) If HOURS = 21.25 and RATE = 3.15, what is the result?

_____

(b) How many decimal places must you reserve for GROSS? _____

_ _ _ _ _ _ _ _ _ _ _ _ _ _ _

(a) 66.9375; (b) four

13. Most computers simply drop, or "chop," digits that overflow reserved spaces. If you had reserved two spaces after the point for GROSS in the last frame, what result would the computer obtain?

_____

_ _ _ _ _ _ _ _ _ _ _ _ _ _

66.93

14. A chopping action by the computer can result in a considerable error when repeated (as in a loop). Rounding can also result in error, but not usually to the same extent. When a programming language has a rounding feature it will act much as you do when figuring by hand. Which of the following explains the rounding?

___(a) All digits to the right of the last desired digit are dropped.

___(b) If the digit immediately to the right of the last desired digit is five or higher, the last desired digit is increased by one and all to the right of it are dropped.

___(c) If the digit immediately to the right of the last desired digit is four or less, all digits to the right of the rightmost desired digit are dropped.

_ _ _ _ _ _ _ _ _ _ _ _ _ _

(b); (c)   (Choice (a) explains chopping, not rounding.)

15. Round each of these floating-point numbers to three decimal places.

(a) 19.1738 _____

(b) 3.14159 _____

(c) 428.934497 _____

(d) 5.22222 _____

_ _ _ _ _ _ _ _ _ _ _ _ _ _

(a) 19.174; (b) 3.142; (c) 428.934; (d) 5.222

16. Which of your answers in frame 15 would be different if you had chopped instead of rounded? _____

_ _ _ _ _ _ _ _ _ _ _ _ _ _

(a); (b)

E NOTATION, OR POWERS OF TEN

Floating-point numbers are frequently written in a form called scientific notation, or E notation. This is especially true when the values are very large or very small. The numbers are written in a special form, and the arithmetic is slightly different, but not very complex.

17. E notation refers to exponents. A floating-point number can be written as a decimal fraction times a power of ten. For example, 17.0 can be written as 0.17×10^2, or 1.7×10^1, or 17.0×10^0.
    Which of the numbers below are written in E notation?

    ___(a) 0.173×10^2
    ___(b) 171
    ___(c) 1.0
    ___(d) 4.72×10^3

    _ _ _ _ _ _ _ _ _ _ _ _ _ _

    (a); (d)

18. The usual, or normalized, form of writing numbers in E notation consists of a decimal fraction with the decimal point just before the first nonzero digit times a power of ten.
    Which of the following numbers is in normalized E-notation form?

    ___(a) 1.03×10^2
    ___(b) 0.006×10^2
    ___(c) 0.671×10^6
    ___(d) 0.4×10^7

    _ _ _ _ _ _ _ _ _ _ _ _ _ _

    (c); (d) (Sometimes another form is called normalized, depending on the installation. We will use the form above.)

19. 1.03×10^2 could be normalized. Then it would be written as 0.103×10^3. When the decimal point is shifted to the left, the exponent:

    ___(a) is decreased
    ___(b) is increased
    ___(c) stays the same

    _ _ _ _ _ _ _ _ _ _ _ _ _ _

    (b)

20. The value of 0.103×10^3 is equal to:

    ___(a) 1.03×10^2
    ___(b) 1030
    ___(c) 103
    ___(d) 0.103×1000

    _ _ _ _ _ _ _ _ _ _ _ _ _ _

    (a); (c); (d)

21. Any number can be written in normalized E notation. Follow the steps below, and answer the questions in converting 4782.6 to E notation.

     As the floating-point number stands, it is equal to 4782.6×10^0. (Remember from Chapter 1 that $10^0 = 1$.)

     (a) First we move the decimal point to just before the first nonzero digit. This gives 0.47826. How many places to the left did we move the point? _____

     (b) Next we increase the power of ten by one for each place to the left the point was moved. What power of ten is included in the answer? _____

     (c) Now write the normalized form of 4782.6: _____

     ― ― ― ― ― ― ― ― ― ― ― ― ― ―

     (a) 4; (b) 10^4; (c) 0.47826×10^4

22. Write the normalized form of each of the following.

     (a) 62.0 _____

     (b) 19.43 _____

     (c) 1.678 _____

     ― ― ― ― ― ― ― ― ― ― ― ― ― ―

     (a) 0.62×10^2; (b) 0.1943×10^2; (c) 0.1678×10^1

23. A number such as 0.006 can also be normalized. Here, however, we move the decimal point to the right, to just in front of the first nonzero digit. And, we <u>decrease</u> the exponent. Since 0.006 is the same as 0.006×10^0, the normalized form is _____.

     ― ― ― ― ― ― ― ― ― ― ― ― ― ―

     0.6×10^{-2}

24. When you move a decimal point to the right, you _____ the exponent. When you move a decimal point to the left, you

     _____ the exponent.

     ― ― ― ― ― ― ― ― ― ― ― ― ― ―

     decrease; increase

25. The following numbers are written in E notation, but are not normalized. Write the normalized form, as in the example.

Example: $476.2 \times 10^2 = 0.4762 \times 10^5$

(a) $1.738 \times 10^3 =$ _____

(b) $0.007 \times 10^1 =$ _____

(c) $0.00008 \times 10^0 =$ _____

(d) $1738.0 \times 10^{-2} =$ _____

– – – – – – – – – – – – – – – –

(a) 0.1738×10^4; (b) 0.7×10^{-1}; (c) 0.8×10^{-4}; (d) 0.1738×10^2 (Remember to count the 10^0 position when going from positive to negative.)

26. Write the following floating-point numbers in normalized E notation.

(a) 27.387 _____

(b) 0.4972 _____

(c) 0.00087 _____

(d) 19783.62 _____

– – – – – – – – – – – – – – –

(a) 0.27387×10^2; (b) 0.4972×10^0; (c) 0.87×10^{-3};
(d) 0.1978362×10^5

27. When computers handle floating-point numbers, they are usually in E notation and normalized. The arithmetic of E notation is fairly straightforward. We will learn a few terms first to simplify discussion of the arithmetic.

Floating-point numbers as stored in a computer will have two parts. The fractional part is called the mantissa. The power of ten is called the exponent. In the number 0.87×10^{-3}:

(a) what is the mantissa? _____

(b) what is the exponent? _____

– – – – – – – – – – – – – – –

(a) 0.87; (b) –3

28. Refer to the examples below and write mantissa or exponent after each of the following.

$$0.92378 \times 10^{-5} \qquad 0.4362 \times 10^4 \qquad 0.3 \times 10^6$$

(a) 0.4362 _____

(b) 4 _____

(c) 0.3 _____

(d) -5 _____

- - - - - - - - - - - - - -

(a) mantissa; (b) exponent; (c) mantissa; (d) exponent

OPERATIONS IN E NOTATION

29. Addition of normalized numbers when the exponents are the same consists of adding the mantissas and using the same exponent. What

   is the sum of 0.372×10^3 and 0.4×10^3? _____

- - - - - - - - - - - - - -

0.772×10^3 $(0.372 + 0.400 = 0.772)$

30. When two normalized numbers do not have the same exponent, one must be adjusted before adding. Suppose you wish to add 0.372×10^3 and 0.4×10^2. You may adjust either one, but let's adjust the second.

   (a) Write 0.4×10^2 with an exponent of 3: _____

   (b) Now add the two mantissas: _____

   (c) Write the sum of 0.372×10^3 and 0.4×10^2: _____

- - - - - - - - - - - - - -

(a) 0.04×10^3; (b) 0.412; (c) 0.412×10^3

31. Suppose we had decided to adjust 0.372×10^3 to use the exponent 2.

   (a) Write 0.372×10^3 with an exponent of 2: _____

   (b) Add that mantissa to 0.4: _____

   (c) What sum do you get for 0.372×10^3 and 0.4×10^2?

   _____

- - - - - - - - - - - - - -

(a) 3.72×10^2; (b) 4.12; (c) 4.12×10^2

32. The computer would now renormalize the result found in frame 31. What is the normalized form of 4.12×10^2? _____

- - - - - - - - - - - - - -

0.412×10^3

33. The answer in frame 31 resulted in <u>overflow</u>. The number overflowed the space because the computer had no space for digits in front of the decimal point. (Even though mathematicians write a zero before the point, computers don't save space for these extra zeros.) For

this reason the number had to be _____.

- - - - - - - - - - - - - -

normalized

34. Suppose a computer can use only eight digits in a mantissa. Some additions may result in more than eight digits. If the computer is set up to chop it will:

____(a) increase the eighth digit by one if the ninth is five or higher
____(b) decrease the eighth digit by one if the ninth is four or lower
____(c) just drop the ninth and later digits
____(d) introduce some error into the result

- - - - - - - - - - - - - - -

(c); (d)

35. Add $.96324117 \times 10^3$ and $.87422221 \times 10^3$.

(a) What is the sum? _____

(b) Normalize it: _____

(c) If it is chopped to eight digits it is: _____

(d) If it is rounded to eight digits it is: _____

- - - - - - - - - - - - - -

(a) 1.83746338×10^3; (b) $.183746338 \times 10^4$; (c) $.18374633 \times 10^4$;
(d) $.18374634 \times 10^4$

36. Neither chopping nor rounding results in an answer that is always accurate. Either one may introduce _____.

- - - - - - - - - - - - - -

error (Rounding never results in a less accurate answer than chopping, although sometimes the answer is the same. But chopping is much more efficient for a computer. If enough digits are used for the numbers, the error introduced will be extremely small. But the programmer must be aware of the possibility.)

37. Subtraction of E-notation numbers, like addition, requires that the two numbers have the same exponent. Solve the E-notation subtractions below.

(a) $0.6 \times 10^2 - 0.423 \times 10^2$

(b) $0.673 \times 10^3 - 0.42 \times 10^1$

- - - - - - - - - - - - - -

(a) 0.177×10^2; (b) 0.6688×10^3

38. Some subtraction problems give results that need to be normalized. Solve the problem below, then normalize the result.

$$0.2521 \times 10^2$$
$$- 0.1867 \times 10^2$$

- - - - - - - - - - - - - -

0.0654×10^2 normalized to 0.654×10^1

39. Multiplication of numbers in E notation does not require that the exponents be the same. You multiply the mantissas and <u>add</u> the exponents.

What is $(0.42 \times 10^2) \times (0.3 \times 10^1)$?

- - - - - - - - - - - - - -

0.126×10^3

40. To multiply E-notation numbers you:

____(a) multiply the exponents
____(b) multiply the mantissas
____(c) add the exponents
____(d) add the mantissas

- - - - - - - - - - - - - -

(b); (c)

41. Solve the following multiplication problems. Normalize the answer if necessary.

(a) $(0.63 \times 10^2) \times (0.71 \times 10^4)$

(b) $(0.1202 \times 10^3) \times (0.4 \times 10^{-2})$

- - - - - - - - - - - - - - -

(a) 0.4473×10^6; (b) 0.04808×10^1, normalized to 0.4808×10^0

42. Division of numbers in E notation involves dividing the mantissas and subtracting the exponent of the denominator from the exponent of the numerator.
      What is the result of the division below?

$$(0.4 \times 10^2) \div (0.2 \times 10^1)$$

- - - - - - - - - - - - - -

2×10^1

43. In the problem below, indicate whether the exponents are added or subtracted.

(a) $(0.278 \times 10^3) \times (0.491 \times 10^{-2})$ _____

(b) $(0.278 \times 10^3) \div (0.91 \times 10^{-1})$ _____

(c) $(0.839 \times 10^3) \div (0.313 \times 10^1)$ _____

- - - - - - - - - - - - - -

(a) add; (b) subtract; (c) subtract

You have learned to do arithmetic with fixed-point numbers, and with numbers written in E notation. You can identify integers. You can write any floating-point or fixed-point number in E notation. Now take the Self-Test on the next page.

SELF-TEST

This Self-Test will help you evaluate whether or not you have mastered the chapter objectives and are ready to go on to the next chapter. Answer each question to the best of your ability. Correct answers are given at the end of the test.

1. Which of the following are integers?

    ___(a) 1.
    ___(b) 37.02
    ___(c) 738
    ___(d) 70

2. Give the result of each of these integer operations.

    (a) 4 + 17 = _____

    (b) 12 − 17 = _____

    (c) 4 x 12 = _____

3. Give the result of each of these integer divisions.

    (a) 17/4 = _____

    (b) 4/17 = _____

    (c) 9/3 = _____

4. Use the formula $I - [(I/J) \times J] = R$ to find the remainder in question 3(a) above.

5. Identify the numbers below as fixed-point or floating-point.

    (a) 17 _____

    (b) 17. _____

    (c) 0.17×10^2 _____

6. Write the numbers below in normalized E notation.

    (a) 0.00038 _____

    (b) 3.1416 _____

    (c) 498762 _____

7. Solve the problems below and normalize the answers.

   (a) $(0.4783 \times 10^2) + (0.8432 \times 10^2)$

   (b) $(0.4783 \times 10^2) + (0.8432 \times 10^1)$

   (c) $(0.4783 \times 10^2) - (0.8432 \times 10^1)$

8. Solve the following problems and normalize the answers. Round to four decimal places.

   (a) $(0.4783 \times 10^2) \times (0.823 \times 10^2)$

   (b) $(0.45 \times 10^7) \times (0.411 \times 10^{-2})$

   (c) $(0.9 \times 10^8) \div (0.3 \times 10^2)$

9. In the number below, circle the mantissa:

   0.98765×10^4

10. (a) If 0.98765×10^4 were rounded to four decimal places, the result would be _____.

    (b) If 0.98765×10^4 were chopped to four decimal places, the result would be _____.

ANSWERS TO SELF-TEST

Compare your answers to the Self-Test with the correct answers given below. If all of your answers are correct, you are ready to go on to the next chapter. If you missed any questions, study the frames indicated in parentheses following the answer.

1. (c); (d)   (frame 1)

2. (a) 21; (b) –5; (c) 48   (frames 1–2)

3. (a) 4; (b) 0; (c) 3   (frame 3)

4. 17 – [(17/4) x 4 = 1]   (frames 4–7)

5. (a) fixed-point; (b) floating-point; (c) floating-point (frames 11–17)

6. (a) 0.38×10^{-3}; (b) 0.31416×10^{1}; (c) 0.498762×10^{6} (frames 21–25)

7. (a) 0.13215×10^{3}; (b) 0.56262×10^{2}; (c) 0.39398×10^{2} (frames 29–32)

8. (a) 0.3936×10^{4}; (b) 0.1850×10^{5}; (c) 0.3×10^{5} (frames 39–42)

9. 0.98765   (frames 27–28)

10. (a) 0.9877×10^{4}; (b) 0.9876×10^{4} (frames 33–36)

CHAPTER SIX

Interest and Mortgage Problems

Some of the problems you will have to solve in computer work will be concerned with percentages. Interest, annuities, and mortgages are basic types of problems you may encounter. This unit is really concerned with basic arithmetic, repeated many times. The concepts are simple—much more so for computers than for you!

When you complete this chapter, you will be able to:

- calculate compound interest for any number of years;

- calculate how long it takes for money to increase to a given amount, at a specific compound rate;

- calculate principal and interest for mortgage payments.

INTEREST

1. Interest is usually stated in the form of "percent per year." If you
    deposit $100 in a bank that pays 5 percent interest per year, at the
    end of the year you would have $105. How is this figure obtained?

    ___(a) multiply 100 by 5
    ___(b) multiply 100 by 1.05
    ___(c) multiply 100 by 0.05 and add the result to 100

    - - - - - - - - - - - - - - -

    (b); (c)

2. In the example in frame 1, interest was calculated only once a year.
    We can say the interest was <u>compounded</u> <u>annually</u>. If interest were
    calculated twice a year, we would say it was:

    ___(a) compounded annually
    ___(b) compounded semiannually
    ___(c) impounded semiannually

    - - - - - - - - - - - - - - -

    (b)

3. When interest is compounded more than once a year, the effective
    rate of interest is higher than the stated rate. In the example in
    frame 1, suppose the interest is to be compounded semiannually, or
    twice a year. Then the rate of interest would be divided in two, one
    for each of the interest periods. The amount on deposit would be
    multiplied by the partial interest, $2\frac{1}{2}$ percent or 0.025. How much

    would you have after the first six-month period? _____

    - - - - - - - - - - - - - - -

    $102.50   [(100 x 0.025) + 100, or 100 x 1.025]

4. The second six-month period the interest rate is the same—2.5 per-
    cent—but the new amount on deposit is larger. How much would you
    have on deposit after one year if the interest is compounded semi-

    annually? _____

    - - - - - - - - - - - - - - -

    $105.06   [(102.50 x 0.025) + 102.50, or 102.50 x 1.025]

5. If a bank compounds interest four times a year, the interest rate is divided by four for the different periods in the year. Suppose a bank is going to calculate interest every two months. Its rate is 6 percent per year. What is the interest percentage for each interest period? _____

- - - - - - - - - - - - - - -

1% (6% divided by 6 times a year gives 1% each time.)

6. Suppose you deposit $200.00 in a bank that pays 6 percent interest, compounded three times a year.

(a) What is the interest rate used in calculations? _____

(b) How much will you have after four months? _____

(c) After eight months? _____

(d) After one year? _____

- - - - - - - - - - - - - - -

(a) 2% (6/3); (b) $204; (c) $208.08; (d) $212.24

7. If the bank in frame 6 compounded interest only once a year, how much would you have after a full year? _____

- - - - - - - - - - - - - -

$212.00

8. A bank pays 6 percent interest compounded four times a year.

(a) What percentage will be used each time interest is calculated?

_____

(b) How many times will interest be calculated in three years?

_____

- - - - - - - - - - - - - -

(a) 1.5%; (b) 12 times

9. Assume a $400 deposit is made in the bank referred to in frame 8. Which of these problems must you solve first?

___(a) 0.06/4
___(b) 400 x 0.015

- - - - - - - - - - - - -

(a)

10. A bank pays 6 percent interest compounded quarterly. A woman deposits $400 and leaves it in the bank for three years. Which of the following flowchart segments could be used to write a program to solve this problem?

____(a)

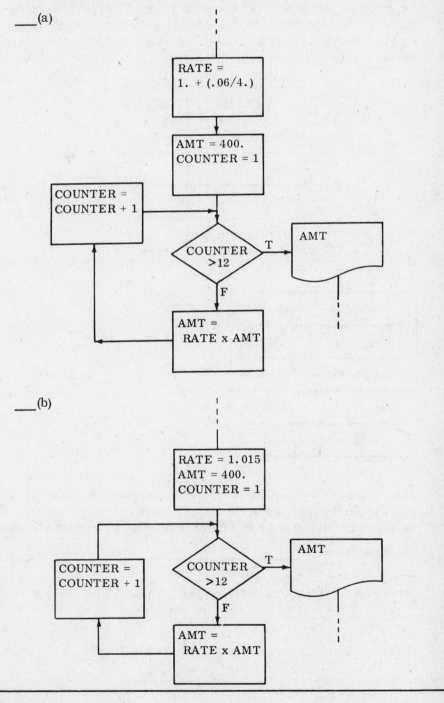

____(b)

- - - - - - - - - - - - - -

Either flowchart will do.

11. To calculate the amount the woman in frame 10 would have on deposit after nine months, consider the block for testing the counter changed to COUNTER > 3. Fill in the values of AMT in the chart below.

PERIOD	AMOUNT
Beginning	$400.00
1st quarter	
2nd quarter	
3rd quarter	

- - - - - - - - - - - - -

PERIOD	AMOUNT
Beginning	$400.00
1st quarter	$406.00
2nd quarter	$412.09
3rd quarter	$418.27

12. Look again at flowchart (b) in frame 10. Assume you want the computer to print every new value of AMT, with the values of COUNTER as shown in the last frame. Where in flowchart (b) would you place

the output block? _____

- - - - - - - - - - - - - -

between AMT = RATE x AMT and COUNTER = COUNTER + 1

13. Each example below gives an annual interest rate and the number of times per year interest is compounded. Give the rate used in each calculation.

(a) 5 percent interest, compounded twice a year _____

(b) 3 percent interest, compounded four times a year _____

(c) 6 percent interest, compounded monthly _____

− − − − − − − − − − − − − − − −

(a) 2.5% (or 0.025); (b) 0.75% (or 0.0075); (c) 0.5% (or 0.005)

14. A bank may pay 5 percent interest, compounded quarterly. Assume you deposit $700 in this bank and leave it six years.

(a) What interest rate will be used each time interest is calculated?

_____

(b) How many times will interest be calculated? _____

− − − − − − − − − − − − − − − −

(a) 1.25% (or 0.0125); (b) 24 (4 x 6)

You have now learned one way to calculate compound interest. You can calculate how much a person will have on deposit after a certain length of time. In Chapter 7, you will learn another way and see how computers solve these problems.

But sometimes you may have a different sort of compound interest problem. Suppose you deposit $500 in a bank that pays 6 percent interest, compounded twice a year. How long will it take until the money doubles? Or until you have $600? We will now consider solutions to this type of problem.

15. Which of the explanations below tells the effect of this flowchart segment?

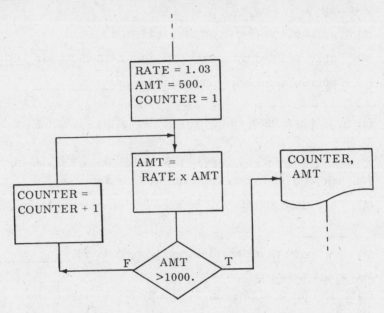

___(a) It will result in a printout of the number of years it will take to double $500 at 6 percent compounded semiannually.

___(b) It will result in a printout of the number of times interest is calculated and added before the original amount doubles.

___(c) It will calculate and print each increase in interest until the amount on deposit has doubled.

— — — — — — — — — — — — — —

(b)

16. The main difference between calculating how much is on deposit after a certain length of time and how long until a certain amount is reached is in:

___(a) how you figure the interest
___(b) how many times a year interest is compounded
___(c) what you test for

— — — — — — — — — — — — — —

(c)

17. Assume you deposit $2000.00 in a bank that pays 5 percent interest compounded semiannually. How much will you have on deposit after two years? (You may find it helpful to construct a chart similar to the one in frame 11 to figure the answer.)

— — — — — — — — — — — — — —

After two years, you will have $2207.62 on deposit.

PERIOD	AMOUNT
1	$2050.00
2	$2101.25
3	$2153.78
4	$2207.62

18. Assume you deposit $10,000 in a savings bank that pays 7 percent interest compounded annually. How many years will it take before you have $15,000?

- - - - - - - - - - - - -

In six years, you will have $15,000.

PERIOD	AMOUNT
1	$10,700.00
2	$11,449.00
3	$12,250.43
4	$13,107.96
5	$14,025.52
6	$15,007.31

19. Assume you deposit $300 every three months into a savings account that pays 5 percent compounded quarterly. At the end of one quarter, how much would you have? _____

- - - - - - - - - - - - -

300 x 1.0125 = 303.75

20. Refer again to the problem in frame 19. How much would you have after one year? Fill in the chart below as you calculate.

PERIOD	BALANCE	DEPOSITS
1	$303.75	$300.00
2		
3		
4		

PERIOD	BALANCE	DEPOSITS
1	$ 303.75	$300.00
2	$ 611.30	$300.00
3	$ 922.69	$300.00
4	$1237.97	$300.00

These problems would all be much easier to program than to calculate. The proper loop in a computer program could manage the calculations and produce printouts at appropriate points. There are other ways, of course, to calculate compound interest. In this section, we have treated the easiest to understand. The same methods, with minor variations, are used in solving problems dealing with annuities, debts, and investments. We will now look at mortgages and see how these are calculated.

MORTGAGES

21. In most mortgage situations, the interest is calculated monthly and figured on the unpaid balance. If the interest is 6 percent, what is the rate per month? _____

0.005, or 0.5%, or $\frac{1}{2}$%

22. Assume a mortgage at 6 percent interest is taken out for $15,000. The monthly payment is $100.00. Each payment is partially for interest and partially for the principal (originally $15,000).

    (a) How much of the first payment is for interest? _____

    (b) How much is for principal? _____

    (a) $75.00; (b) $25.00

23. The interest amount of the second payment is calculated on the new principal balance. After the first payment, the balance of the mortgage in frame 22 is _____.

- - - - - - - - - - - - - - -

$14,975.00

24. Use the information in the last two frames to answer these questions.

    (a) How much interest is included in the second payment?

    _____.

    (b) How much principal is paid in the second payment?

    _____

    (c) What is the unpaid balance after the second payment?

    _____

- - - - - - - - - - - - - - -

    (a) $74.88 (14,975 x 0.005); (b) $25.12 (100.00 − 74.88);
    (c) $14,949.88 (14,975.00 − 25.12)

25. The chart below reflects the calculations you have performed in the last three frames. Complete the chart now for payments 3 and 4.

PAYMENT	INTEREST	PRINCIPAL	BALANCE
1	$75.00	$25.00	$14,975.00
2	$74.88	$25.12	$14,949.88
3			
4			

- - - - - - - - - - - - - -

PAYMENT	INTEREST	PRINCIPAL	BALANCE
1	$75.00	$25.00	$14,975.00
2	$74.88	$25.12	$14,949.88
3	$74.75	$25.25	$14,924.63
4	$74.62	$25.38	$14,899.25

26. Suppose a mortgage on a small house is $9,950. The rate is 6 percent and the monthly payment is $59.70. The house will be completely paid for in thirty years.

(a) How many payments will the purchasers make? _____

(b) How many times will interest be calculated? _____

- - - - - - - - - - - - - -

(a) 360; (b) 360

27. Fill in the chart below for the first three payments on the mortgage described in frame 26.

PAYMENT	INTEREST	PRINCIPAL	BALANCE
1			
2			
3			

- - - - - - - - - - - -

PAYMENT	INTEREST	PRINCIPAL	BALANCE
1	$49.75	$ 9.95	$9,940.05
2	$49.70	$10.00	$9,930.05
3	$49.65	$10.05	$9,920.00

28. The chart below gives interest, principal, and balance for payments 357 and 358 of the same mortgage. Fill in the figures for the final two payments. What is the amount of the last payment?

PAYMENT	INTEREST	PRINCIPAL	BALANCE
357	$0.96	$58.74	$132.95
358	$0.66	$59.04	$ 73.91
359			
360			

The last payment is _____.

- - - - - - - - - - - - - -

PAYMENT	INTEREST	PRINCIPAL	BALANCE
357	$0.96	$58.74	$132.95
358	$0.66	$59.04	$ 73.91
359	$0.37	$59.33	$ 14.58
360	$0.07	$14.58	-0-

The last payment is $14.65.

29. Assume you need to prepare a mortgage table showing payment numbers, interest, principal, and balance. The first balance printed
will be after the first payment. The last balance printed will be
zero. The input will be annual interest rate, amount of mortgage,
and amount of each monthly payment. Which of the following flowcharts might be used to solve this problem?

___(a)

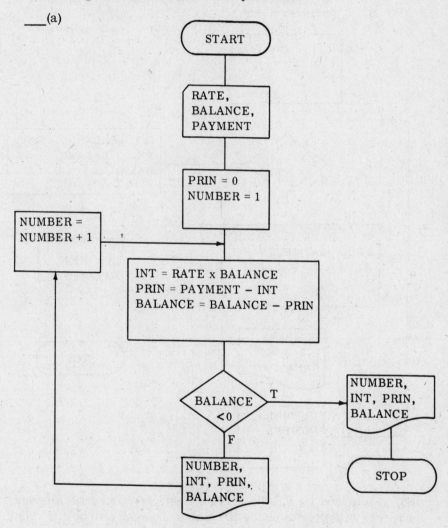

(continued on next page)

___(b)

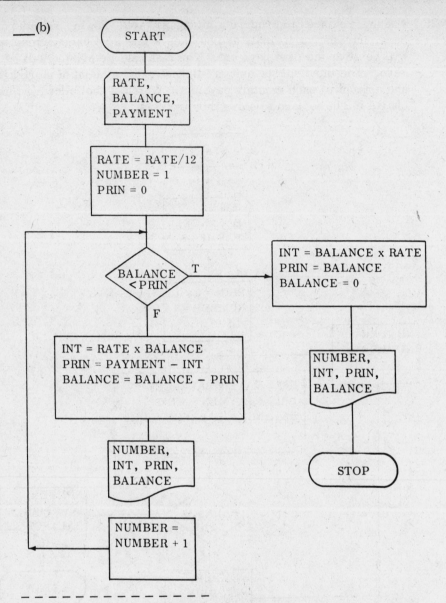

START

RATE,
BALANCE,
PAYMENT

RATE = RATE/12
NUMBER = 1
PRIN = 0

BALANCE < PRIN

INT = BALANCE x RATE
PRIN = BALANCE
BALANCE = 0

INT = RATE x BALANCE
PRIN = PAYMENT − INT
BALANCE = BALANCE − PRIN

NUMBER,
INT, PRIN,
BALANCE

NUMBER,
INT, PRIN,
BALANCE

STOP

NUMBER =
NUMBER + 1

_ _ _ _ _ _ _ _ _ _ _ _ _ _ _

(b) (Flowchart (a) fails in several respects. The most obvious error is in failing to divide the annual interest rate by 12. The last balance printed will be less than zero, and the calculation of the last interest will result in a negative number.)

30. Suppose your input for the mortgage table in frame 29 set the rate at 6.75 percent, the original mortgage balance at $15,000, and the payments at $120 a month. What interest rate (to six decimal places) is used in calculating each interest amount? _____ What is the principal balance after the first payment? _____

- - - - - - - - - - - - - - - -

   0.005625; $14,964.38

31. Suppose the input information for flowchart (b) in frame 29 is RATE = 7%, BALANCE = $18,000, and PAYMENT = $130.00. What would be the value of BALANCE after the second payment?

   _____

- - - - - - - - - - - - - - - -

   $17,948.65 (rate/month is 0.0058)

You have learned the basics of solving interest and mortgage problems—by hand and with computers. After calculating just a few interest and mortgage problems, you probably appreciate how calculators and computers simplify life in the business world. The time saved and the accuracy achieved are tremendous in just this area alone.

   You can now review the material presented in this unit by solving these review problems before taking the Self-Test.

REVIEW FRAMES

32. Suppose your bank pays $4\frac{1}{2}$ percent interest compounded semiannually. You receive a $1000 windfall and deposit it. How much will you have after two years?

- - - - - - - - - - - - - - - -

After two years you will have $1093.08.

PERIOD	BALANCE
1	$1022.50
2	$1045.51
3	$1069.03
4	$1093.08

33. Suppose an investment company guarantees you 8 percent annual return on your investment. You give them $8000. How long will it take for you to have $10,000?

- - - - - - - - - - - - - -

It will take 3 years for you to have $10,000.

PERIOD	BALANCE
1	$ 8,640.00
2	$ 9,331.20
3	$10,077.70

34. Suppose you want to buy a house for $24,500. The interest is 9 percent and the monthly payments are $225.00. What is the balance after the second payment?

- - - - - - - - - - - - - -

$24,417.19

PAYMENT	INTEREST	PRINCIPAL	BALANCE
1	$183.75	$41.25	$24,458.75
2	$183.44	$41.56	$24,417.19

(Interest rate is 0.0075 per month.)

SELF-TEST

This Self-Test will help you evaluate whether or not you have mastered the chapter objectives and are ready to go on to the next chapter. Answer each question to the best of your ability. Correct answers are given at the end of the test.

1. Assume a bank pays $4\frac{1}{2}$ percent interest, compounded semiannually. You deposit $700.00. How much will you have in the bank after one year, if you make no withdrawals?

2. A bank pays 4 percent interest, compounded quarterly. You deposit $200.00 on the first banking day of each year. How much will you have in the bank after two full years?

3. At 6 percent compounded annually, how many years will it take for $100 to be worth $150?

4. A couple takes out a $19,500 mortgage. The interest rate is 6 percent. The monthly payment is $128.53. What is the balance after the second payment?

ANSWERS TO SELF-TEST

Compare your answers to the Self-Test with the correct answers given below. If all of your answers are correct, you are ready to go on to the next chapter. If you missed any questions, study the frames indicated in parentheses following the answer.

1. $731.85

PERIOD	BALANCE
1	$715.75
2	$731.85

Interest rate = 0.0225 per period   (frames 1-5)

2. $424.68

PERIOD	DEPOSIT	BALANCE
1	$200.00	$202.00
2		$204.02
3		$206.06
4		$208.12
5	$200.00	$412.20
6		$416.32
7		$420.48
8		$424.69

Interest rate = 0.01 per period   (frames 19-20)

3. 7 years
    ($150.37)

YEAR	BALANCE
1	$106.00
2	$112.36
3	$119.10
4	$126.25
5	$133.83
6	$141.86
7	$150.37

(frames 15–18)

4. $19,437.78

PAYMENT	INTEREST	PRINCIPAL	BALANCE
1	$97.50	$31.03	$19,468.97
2	$97.34	$31.19	$19,437.78

(frames 21–28)

CHAPTER SEVEN
Sequences and Series

This chapter introduces some of the concepts and computations involved in sequences and series as they are used in mathematics. Sequences and series simplify many business and data problems including some similar to those you struggled with in the last chapter. You will see how sequences and series differ and learn the relationships between them. You will learn a few of their properties and the concept of the limit. Algorithms for finding sums of series are given in flowchart form.

When you complete this chapter, you will be able to:

- identify examples of sequences and series as finite or infinite;

- construct a sequence or series, given a general term;

- use the concept of the limit in deciding if a series has a sum;

- specify whether a series diverges or converges, given facts about the limit of its sum;

- read and interpret a flowchart for calculating the sum of a series.

Your work in this and subsequent chapters will be facilitated if you already have a knowledge of basic arithmetic including operations with fractions and decimals. If you feel you need a review in this area, you might use another Self-Teaching Guide as a reference: Math Shortcuts by Flora M. Locke (New York: John Wiley & Sons, 1972)

SEQUENCES

Sequences and series are terms you use in everyday activities, but probably not in the mathematical sense. You may speak of a series of bad
breaks, for example. In the next several frames, you will learn how
these terms are used in the language of mathematics.

1. A <u>sequence</u> is a set of ordered elements, called terms, in which
there is a specific first term, a second term, and so on, including
an <u>n</u>th term. Here n represents any positive integer; there may be
any number of positive terms.
    Which of the examples below might be a sequence?

___(a) 1, 2, 3

___(b) $\frac{1}{2}, \frac{1}{4}, \frac{1}{8}$

___(c) the set of all integers in numerical order

_ _ _ _ _ _ _ _ _ _ _ _ _

(a); (b) (There is no specific first term for the set of all integers.)

2. A sequence may be finite or infinite. If it is finite, the terms can
be counted. If it is infinite, the terms are too many to be counted.
They cannot all be written down.
    Write finite or infinite for the sequences described below.

(a) the sequence of all positive integers _____

(b) the sequence of all integers between –12 and +738

_____

_ _ _ _ _ _ _ _ _ _ _ _ _

(a) infinite (You would never reach the "largest" integer.)
(b) finite

3. An infinite sequence cannot, of course, be completely written out.
It can, however, be written as shown below, where t(1) means term
1, and t(n) means the <u>n</u>th term:

$$t(1), t(2), t(3), \ldots , t(n), \ldots$$

The symbol ". . ." means something is left out. It may be one term,
seventy-five thousand terms, or an infinite number. The ". . ." at
the end means the rest of the sequence is omitted, from n to infinity.
    Suppose you are considering the infinite sequence beginning
$\frac{1}{1}, \frac{1}{2}, \frac{1}{3}, \ldots$ Here the <u>n</u>th term would be $\frac{1}{n}$. What could you write
for the <u>n</u>th term if n = 12? _____

- - - - - - - - - - - -

$$\frac{1}{12}$$

4. In the problem in frame 3, what would the n̲th term be if n = 912?

    _____ What would you write if you want the n̲th term to be used

    for any value of n? _____

- - - - - - - - - - - -

$$\frac{1}{912}; \frac{1}{n}$$

SERIES

5. A sequence is a set of terms in a specific order. A series is an ex-
    pression formed from a sequence, in which the terms of the sequence
    are added together, as in the general form below:

    $$t(1) + t(2) + t(3) + \ldots + t(n) + \ldots$$

    Write sequence or series for each of the items listed below.

    (a) $\frac{1}{1} + \frac{1}{2} + \frac{1}{3} + \frac{1}{4}$ _____

    (b) 1, 1, 2, 3, 5, . . .   _____

    (c) t(1), t(2), t(3) _____

- - - - - - - - - - - -

(a) series; (b) sequence (Each term is the sum of the two preceding
terms: 1 + 1 = 2, 1 + 2 = 3, 2 + 3 = 5, etc.); (c) sequence
(Notice that the critical difference is whether or not the terms are
added together to form an expression.)

6. Like a sequence, a series can be infinite or finite. Which notation
    below might represent an infinite series?

    ___(a) $\frac{1}{1} + \frac{1}{2} + \frac{1}{3} + \ldots + \frac{1}{n} + \ldots$

    ___(b) $t(1) + t(2) + t(3) + \ldots + t(n) + \ldots$

    ___(c) 1, 1, 2, 3, 5, 8, . . .

- - - - - - - - - - - -

(a); (b) (Choice (c) represents an infinite sequence.)

7. The nth term of a sequence or series can be given as an expression using n, whatever value n has. For example, the nth term in the sequence of all positive integers is simply n. If n = 93, then the nth term is 93.

   Consider the sequence below:

   $$\frac{1}{2}, \frac{1}{4}, \frac{1}{6}, \ldots, \frac{1}{2n}, \ldots$$

   Suppose, in this case, n = 40. What is the value of the nth term of the sequence? _____

   ‒ ‒ ‒ ‒ ‒ ‒ ‒ ‒ ‒ ‒ ‒ ‒ ‒ ‒

   $\frac{1}{80}$ (The 40th term in the sequence is $\frac{1}{2(40)}$.)

8. Suppose the nth term of a sequence is given by the expression $\frac{1}{2^n}$. The first term is given below. Write the next three terms of the sequence.

   $$\frac{1}{2^1} = \frac{1}{2}, \underline{\hspace{3cm}}, \underline{\hspace{3cm}}, \underline{\hspace{3cm}}$$

   ‒ ‒ ‒ ‒ ‒ ‒ ‒ ‒ ‒ ‒ ‒ ‒ ‒ ‒

   $\frac{1}{2^2} = \frac{1}{4}, \frac{1}{2^3} = \frac{1}{8}, \frac{1}{2^4} = \frac{1}{16}$

SUMS OF SERIES

9. We have said that a series can be represented by:

   $$t(1) + t(2) + t(3) + \ldots + t(n) + \ldots$$

   It is clear then that we can talk about a <u>sum</u> of a series. Two finite series are given below. Find each sum.

   (a) $\frac{1}{1} + \frac{1}{10} + \frac{1}{100} + \frac{1}{1000} = $ _____

   (b) $1 + 2 + 3 + 4 + 5 + 6 = $ _____

   ‒ ‒ ‒ ‒ ‒ ‒ ‒ ‒ ‒ ‒ ‒ ‒ ‒

   (a) $\frac{1111}{1000}$ (or 1.111); (b) 21

10. The sum of an infinite series is a different problem, since not even a computer can actually add up an infinite number of terms. But there are three general possibilities. One is that the sum might become infinitely larger as n gets larger and larger, or closer to infinity (symbolized $n \to \infty$). The sum might get closer and closer to, but never actually reach, some finite number. Or the sum might not exist at all. Consider the series below:

$$1 + 2 + 3 + \ldots + n + \ldots$$

The sum of this series is probably:

___(a) finite
___(b) infinite

- - - - - - - - - - - - - - -

(b)

11. One way to find the sum of a series is to find several "partial sums." To do this add just the first two terms, then the first three, then the first four, etc. Then you compare these partial sums to see if a pattern exists.

   Examine the series below, then calculate the partial sums as indicated.

$$\frac{1}{2} + \frac{1}{4} + \frac{1}{8} + \ldots + \frac{1}{2^n} + \ldots$$

(a) The sum of t(1) and t(2) is _____.

(b) The sum of t(1), t(2), and t(3) is _____.

(c) The sum of t(1), t(2), t(3), and t(4) is _____.

- - - - - - - - - - - - - - -

$\frac{3}{4}$; $\frac{7}{8}$; $\frac{15}{16}$

12. If we use the same denominator for the three partial sums in frame 11, we have $\frac{12}{16}$, $\frac{14}{16}$, and $\frac{15}{16}$. Which statement below describes the pattern of these partial sums?

___(a) They are decreasing as n increases.
___(b) They are increasing as n increases.

- - - - - - - - - - - - - - -

(b)

LIMITS OF SUMS

13. Sometimes the sum of an infinite series is finite. We call that sum the <u>limit</u> of the sum of the series. It is the number that the sum approaches, or gets extremely close to, as n approaches infinity.

   Below is a list of the first eight partial sums for the problem of the last two frames:

$$\frac{3}{4}, \frac{7}{8}, \frac{15}{16}, \frac{31}{32}, \frac{63}{64}, \frac{127}{128}, \frac{255}{256}, \frac{511}{512}$$

The limit of the sequence of numbers representing the partial sums of the infinite series might be:

___(a) 0
___(b) 1
___(c) ∞

- - - - - - - - - - - - - -

(b) (The partial sums are getting very close to 1, but will never pass 1.)

14. Consider an infinite series whose <u>n</u>th term is $\frac{1}{2^{(n-1)}}$. Write the first four terms of this series.

_____ _____ _____ _____

- - - - - - - - - - - - - -

$1 + \dfrac{1}{2} + \dfrac{1}{4} + \dfrac{1}{8} + \dfrac{1}{2^{(1-1)}} = \dfrac{1}{2^0} = \dfrac{1}{1}$ Remember from Chapters 1

and 2 that any number to the zero power equals 1 ($n^0 = 1$).

15. Compare the infinite series of the last frame to the one in frames 11 through 13. The sum of the series in frame 14 is probably:

___(a) the same as the one in frame 11
___(b) larger than the one in frame 11
___(c) smaller than the one in frame 11

- - - - - - - - - - - - - -

(b) (Since the first term is 1, the sum must be larger than that in frame 11, which approaches 1.)

16. The limit of the sum of an infinite series is symbolized by:

$$\lim_{n \to \infty}$$

This is read "the limit as n approaches infinity." If we use S to represent the sum, we might write:

$$S = \lim_{n \to \infty} (1 + \frac{1}{2} + \frac{1}{4} + \frac{1}{8} + \ldots + \frac{1}{2^{(n-1)}})$$

Compare this series to the one from frame 11:

$$\frac{1}{2} + \frac{1}{4} + \frac{1}{8} + \ldots + \frac{1}{2^n}$$

The limit of the sum of the new series is _____.

- - - - - - - - - - - - - - -

2 (This series contains all the terms in the series of frame 11, plus the term 1.)

17. When the limit of a sum of a series exists and is finite, the series is said to <u>converge</u>. If the limit doesn't exist, or is infinite (as in 2 + 4 + 6 + . . .), the series <u>diverges</u>.
    Match the descriptions and terms below.

___(a) Series has an infinite limit.
___(b) Series has a finite limit.   1. converges
___(c) Series has no limit.   2. diverges

- - - - - - - - - - - - - - -

(a) 2; (b) 1; (c) 2

18. We have discussed the sums of the series given below. Write converges or diverges after each.

(a) $\frac{1}{2} + \frac{1}{4} + \frac{1}{8} + \frac{1}{16} + \ldots$ _____

(b) $1 + \frac{1}{2} + \frac{1}{4} + \frac{1}{8} + \ldots$ _____

(c) $1 + 2 + 3 + 4 + \ldots$ _____

- - - - - - - - - - - - - - -

(a) converges; (b) converges; (c) diverges

APPLICATIONS

This chapter has been fairly simple mathematics. But of what use is it to you? Many problems in science involve infinite series. Probability and statistics, too, involve many of the features of series and sequences.

We have restricted ourselves to a few very easily added series that are fairly definite in their outcomes. Now we shall consider the mechanisms involved in finding sums to a specified accuracy. Computers can handle these problems very efficiently, as you shall discover in the remainder of this chapter.

19. Suppose you deposit $500 in a bank and receive 4 percent interest, compounded semiannually. You plan to leave it in the bank for three years. In the last chapter, you learned one way to calculate how much money would be on deposit. The amount on deposit after each interest period can also be given by the expression:

$$P_n = P(1 + r)^n$$

(The mathematical proof of this will not be given here.) Thus, after the first interest period you would have $P_1 = 500(1.02)^1$ on deposit. After the second period you would have $P_2 = 500(1.02)^2$, and after the sixth period you would have $P_6 = 500(1.02)^6$ to withdraw. Do these six terms constitute a sequence or a series in this calculation?

_____

– – – – – – – – – – – – – –

sequence (They are not added. The last term, P_6, represents the amount on deposit after six interest periods.)

20. Suppose now that the bank and initial deposit are the same as in frame 19. But at the beginning of each interest period you deposit $500. Now the amount to be withdrawn after three years will be considerably larger. You have the original $500 deposit in for six interest periods, the next $500 in for five periods, the next for four, and so on until the last $500 is on deposit for only one interest period. The amount withdrawn will now be expressed as:

___(a) an infinite sequence
___(b) a finite sequence
___(c) an infinite series
___(d) a finite series

– – – – – – – – – – – – – –

(d)

21. Using as the <u>n</u>th term $P(1 + r)^n$, write the entire series for solving the problem posed in frame 20. Write the terms only—don't expand them or add them.

$500(1.02)^1 +$ _____ + _____ + _____ + _____ +

_____

- - - - - - - - - - - - - - -

$500(1.02)^1 + 500(1.02)^2 + 500(1.02)^3 + 500(1.02)^4 + 500(1.02)^5 + 500(1.02)^6$

22. The flowchart segment below shows the steps a computer might go through to sum the series of frame 21.

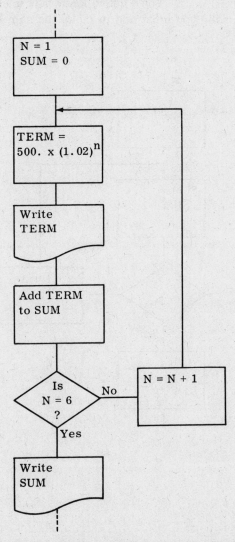

Study the flowchart on page 169, then answer the following questions.

(a) How many separate times will information be written? _____

(b) How would you change the flowchart if you want to know only the amount on deposit after three years, and not the amount after each calculation? _____

– – – – – – – – – – – – – – –

(a) seven (six TERMs and one SUM)
(b) remove the block for Write TERM

23. Suppose you are interested in adding consecutive positive integers $(1 + 2 + 3 + 4 + \ldots)$. You want to know how many terms, or integers in this case, you must add to make the sum be over 1000.

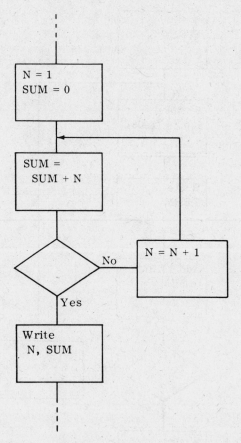

Which of the expressions below might you write in the decision block of the flowchart on page 170?

___(a) Is SUM < 1000?

___(b) Is N > 1000?

___(c) Is SUM > 1000?

___(d) Is SUM − 1000 > 0?

— — — — — — — — — — — — — — —

(c); (d) (These two are equivalent; both check to see if the total is greater than 1000.)

24. Suppose you know the formula for the nth term in an infinite series. You also know that the series converges. Which of the following are true statements?

___(a) The ninth term will be smaller than the tenth term.

___(b) The sum has a finite limit.

___(c) The sum of the first n terms approaches a specific amount as n approaches infinity.

— — — — — — — — — — — — — — —

(b); (c) (The terms will always decrease in magnitude as n gets larger if the series is going to converge.)

25. Study the flowchart segment below.

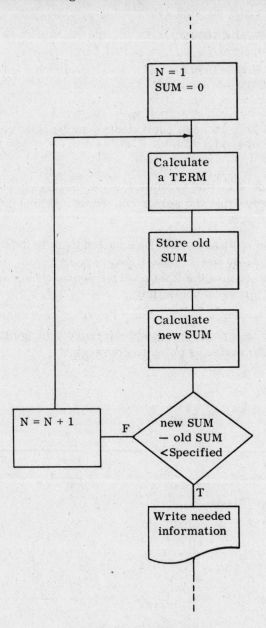

This segment gives the steps in calculating:

___(a) the sum of a series until the last term added makes as little difference as desired

___(b) the sum of a series until the last term added makes no difference at all in the total

___(c) the limit of a series

- - - - - - - - - - - - - -

(a) (The last term will always make a little difference, even though the amount may be very small.)

REVIEW FRAMES

26. Write series or sequence after each of the following.

(a) $1 + \dfrac{1}{2} + \dfrac{1}{3} + \dfrac{1}{4} + . . . + \dfrac{1}{n} + . . .$ _____

(b) $2 + 4 + 6$ _____

(c) $\dfrac{1}{9}, \dfrac{1}{8}, \dfrac{1}{7}, \dfrac{1}{6}, \dfrac{1}{5}, \dfrac{1}{4}, \dfrac{1}{3}, \dfrac{1}{2}$ _____

(d) $1, 2, 3, 4, 5, 6, . . .$ _____

- - - - - - - - - - - - - -

(a) series; (b) series; (c) sequence; (d) sequence

27. Which of the series and sequences of the preceding frame are infinite? _____

- - - - - - - - - - - - - -

(a); (d)

28. If the sum of a series approaches a finite limit, we say the series:

___(a) is infinite

___(b) converges

___(c) diverges

- - - - - - - - - - - - - -

(b)

29. The sum of the series $1 + 1 + 1 + 1 + . . .$ approaches an infinite limit of n as n approaches infinity. Does this series converge or diverge? _____

- - - - - - - - - - - - - -

diverge

30. Consider a series whose nth term is $\dfrac{1}{3^n}$. Write the first three terms of this series.

_____ _____ _____

- - - - - - - - - - - - - -

$$\frac{1}{3} + \frac{1}{9} + \frac{1}{27}$$

31. Suppose you want a computer to calculate partial sums for the infinite series given by $\frac{1}{3^n}$ until the difference between the last two partial sums is less than 0.00001.

    In the following flowchart segments:

    N is a counter, which represents the term;

    SUM is the current partial sum;

    OLDSUM is the previous partial sum (this must be saved for comparing with the current one).

    Which of the flowchart segments might you follow?

___(a)   ___(b)

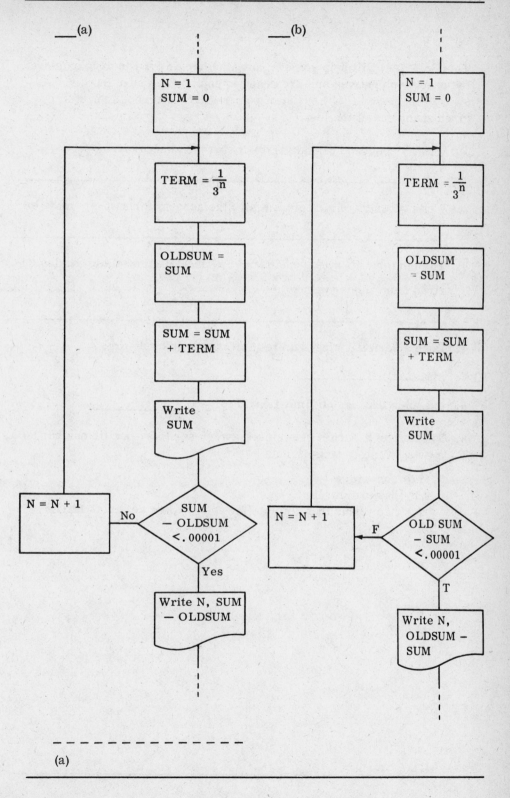

(a)

SELF-TEST

This Self-Test will help you evaluate whether or not you have mastered the chapter objectives and are ready to go on to the next chapter. Answer each question to the best of your ability. Correct answers are given at the end of the test.

1. Write a sequence of eight terms using the general term $\frac{1}{2n}$.

   _____

2. Write a series of six terms where the general form of a term is 4n.

   _____

3. The $\underline{n}$th term in a sequence is $\frac{1}{n + 6}$. Indicate the infinite series derived from this sequence.

   _____

4. Does a series having a finite limit converge or diverge?

   _____

5. A series that has no limit is said to _____.

6. The sum of a series gets closer and closer to a specific number as n gets infinitely large. This series:

   ___(a) converges
   ___(b) has a finite limit
   ___(c) has terms that get smaller and smaller as n gets larger

ANSWERS TO SELF-TEST

Compare your answers to the Self-Test with the correct answers given below. If all of your answers are correct, you are ready to go on to the next chapter. If you missed any questions, study the frames indicated in parentheses following the answer.

1. $\dfrac{1}{2}, \dfrac{1}{4}, \dfrac{1}{6}, \dfrac{1}{8}, \dfrac{1}{10}, \dfrac{1}{12}, \dfrac{1}{14}, \dfrac{1}{16}$ (frames 1-4)

2. $4 + 8 + 12 + 16 + 20 + 24$ (frames 5-7)

3. $\dfrac{1}{7} + \dfrac{1}{8} + \dfrac{1}{9} + \ldots + \dfrac{1}{n + 6} + \ldots$ (frames 3-9)

4. converge (frames 17-18)

5. diverge (frames 17-18)

6. (a); (b); (c) (frames 15-18)

CHAPTER EIGHT
Probability

Much of what we do in life depends on our expectations for the future. What will happen tomorrow? Next week? Next year? This is true of everyday decisions. Shall I wash the car or will it rain tomorrow? It is also true in business, psychology, and the broad field of computer applications. The details of the expectations will vary, but the basis is the same—probability theory. Probability plays a part in every aspect of this complex world we live in. Computers are vital in figuring the probabilities for such difficult events as stock market changes, armament plans, and components in a space satellite.

In this chapter we will consider basic probability theory as it applies to many events. The probabilities of various combinations and permutations will also be computed. When you complete this chapter, you will be able to:

- calculate probabilities of simple events;
- calculate probabilities of composite events;
- calculate $\binom{N}{S}$ for finding the number of combinations;
- calculate $(N)_S$ for finding the number of permutations;
- calculate probabilities of certain combinations and permutations.

Further information on probability theory can be found in the following books:

Hodges, Jr., J. L. and E. L. Lehmann, Basic Concepts of Probability and Statistics (San Francisco: Holden-Day, Inc., 1964)

Kemeny, Schleifer, Snell, and Thompson, Finite Mathematics with Business Applications (Englewood Cliffs, N.Y.: Prentice-Hall, Inc., 1962)

Koosis, Donald J., Probability (a Wiley Self-Teaching Guide) (New York: John Wiley & Sons, 1972)

PROBABILITIES OF SIMPLE EVENTS

Probability theory had its beginnings in the study of gambling devices such as dice. Even today some of the clearest examples of simple probability involve dice, coins, and decks of cards. But the principles you will learn here carry far beyond these simple games. You will be able to apply this knowledge to your everyday activities, your work, and your studies.

1. An ordinary coin has two sides. If you flip it, there are two possibilities: heads will come up or tails will come up. These two results are equally likely to occur in a "fair" coin. Suppose you flip a coin fifty times. You would expect to get about how many heads?

   _____

   – – – – – – – – – – – – – –

   about 25

2. In flipping a coin we call the occurrence of a head (H) a simple event. Likewise, the occurrence of a tail (T) is a simple event. A <u>simple event</u> means a simple result. All of the simple results possible from an action (or experiment) make up an <u>event set</u>. Or, put another way, an event set is <u>all</u> the possibilities. What is the

   event set for flipping a coin? _____

   – – – – – – – – – – – – – – – –

   heads and tails (A set is simply a collection of things; they could be objects, events, or ideas.)

3. We will abbreviate simple event as e. We will use $\mathcal{E}$ to represent the event set (all the different simple events possible). Suppose you are considering the simple event H (the occurrence of heads).

   (a) What does e refer to? _____

   (b) How many events would be in $\mathcal{E}$? _____

   – – – – – – – – – – – – – – –

   (a) H; (b) 2

4. The probability of a simple event is abbreviated p(e). Write the

   abbreviation for the probability of heads. _____

   – – – – – – – – – – – – – – –

   p(H) (or p(heads), though the shorter symbol is usually used)

5. What is the probability of getting heads when you flip a coin? You might say 50-50. Actually, however, a probability is always between 0 and 1, inclusive. The term 50-50 means $\frac{1}{2}$, so p(H) is $\frac{1}{2}$ or 0.5. If p(e) = 0, the event will never occur. If p(e) = 1, the event will always occur. Most probabilities, in practice, lie between 0 and 1.

Which of the statements below is correct for every p(e)?

____(a) $0 \gtrless p(e) \gtrless 1$
____(b) $0 < p(e) < 1$
____(c) $p(e) \neq 0$
____(d) $0 \leq p(e) \leq 1$

- - - - - - - - - - - - - - -

(d)

6. If events are equally likely, they have the same probability. The probabilities of all simple events in an event set add up to 1. Consider our coin-flipping experiment.

(a) How many simple events (e) are in the event set $\mathcal{E}$? _____

(b) If the probabilities add up to 1, what is p(H)? _____

(c) What is p(T)? _____

- - - - - - - - - - - - - - -

(a) 2; (b) $\frac{1}{2}$ (or 0.5); (c) $\frac{1}{2}$ (or 0.5)

7. A fair die has six faces, all of which are equally likely to come up.

(a) What are two of the simple events that could occur when you toss the die? _____

(b) How many members of $\mathcal{E}$ are there? _____

(c) What is the sum of the probabilities of all simple events? _____

- - - - - - - - - - - - - - -

(a) 1, 2, 3, 4, 5, 6 (any two); (b) 6 (Six different simple events are possible.); (c) 1

8. Based on your answers to the questions in frame 7, what is p(3) when you toss a fair die? _____

_ _ _ _ _ _ _ _ _ _ _ _ _ _
1/6

9. The formula for calculating probability in equally likely cases is:

$$p(e) = \frac{\# e}{\# \mathcal{E}}$$

This means that the number of ways the simple event can occur is divided by the number of simple events possible. When we say $p(3) = 1/6$ we mean the number of ways the event "3" can occur is

_____, and the number of elements in the event set is _____.

_ _ _ _ _ _ _ _ _ _ _ _ _ _
1; 6

10. Consider a standard 52-card deck of cards. If the deck is well shuffled, the probability of drawing any specific card is 1/52. There is only one way any specific card can be drawn. But what is p(A), or the probability of drawing an ace?

(a) How many ways could an ace be drawn (or how many aces are available)? _____

(b) What is the number of elements in the event set? _____

(c) What is p(A) if $p(e) = \frac{\# e}{\# \mathcal{E}}$? _____

_ _ _ _ _ _ _ _ _ _ _ _ _ _
(a) 4; (b) 52; (c) 4/52 or 1/13

11. Consider again the well-shuffled deck of 52 cards. Remember that $p(e) = \frac{\# e}{\# \mathcal{E}}$. Find the probabilities of the simple events listed below.

(a) a black two _____

(b) a heart _____

(c) a nine _____

_ _ _ _ _ _ _ _ _ _ _ _ _ _
(a) p(black 2) = 2/52 = 1/26; (b) p(♥) = 13/52 = 1/4;
(c) p(9) = 4/52 = 1/13

12. In our well-shuffled deck, what is the sum of p(♦), p(♥), p(♣), and p(♠)? _____

----

1 (You would always draw either a diamond, a heart, a club, or a spade. Therefore the probability is 1.)

PROBABILITIES OF COMPOSITE EVENTS

13. You can figure out that p(red card) is $\frac{1}{2}$. You could calculate this by adding p(♥) and p (♦): $\frac{1}{4} + \frac{1}{4} = \frac{1}{2}$. This can be considered a <u>composite event</u> (E). The probability of a composite event—p(E)—is the sum of the simple events that make up the composite events. Suppose the composite event consists of the simple events 5 and 6 for a fair die toss.

(a) p(5) = _____

(b) p(6) = _____

(c) p(E) = _____

(a) 1/6; (b) 1/6; (c) 1/3 (1/6 + 1/6)
NOTE: The composite event in frame 13 deals with <u>mutually</u> <u>exclusive</u> events. Drawing a heart and drawing a diamond are mutually exclusive; you can't get both on a single draw. Another composite event might be drawing a heart or drawing a king; these are <u>not</u> mutually exclusive events since you can draw both a heart and a king— the king of hearts—on one draw. To find p(E) in this case, add the two probabilities and then subtract the probability of the overlapping events: p(♥) = 13/52, p(K) = 4/52, and p(K♥) = 1/52; thus, p(♥ ∨ K) = 13/52 + 4/52 − 1/52 = 16/52 = 4/13.

14. Suppose you want to know the probability of getting an odd number when you toss a fair die.

(a) What does the event set consist of? _____

(b) What does the composite event consist of? _____

(c) What is p(E)? _____

(a) 1, 2, 3, 4, 5, 6 (Six simple events; odd or even is not a <u>simple</u> event.); (b) 1, 3, 5 (3 simple events); (c) 1/2 (3/6)

15. What is the probability that you will draw either an ace or a face card (K, Q, or J) from a well-shuffled deck? _____

– – – – – – – – – – – – – – –

16/52 or 4/13

16. The <u>complement</u> of a composite event is made up of all possible single events that are not in E. The complement of the composite event in frame 15 includes cards 2, 3, 4, 5, 6, 7, 8, 9, and 10 in all four suits. What is the complement of the composite event "an even number up on a die toss"? _____

– – – – – – – – – – – – – – –

an odd number up (or 1, 3, and 5)

17. The composite event for tossing a fair coin is H.

   (a) What is the complement? _____

   (b) What is the probability of the complement? _____

– – – – – – – – – – – – – – –

(a) T (all possible simple events not in E); (b) 1/2

18. In the last frame, the probability of E is 1/2. The probability of the complement of E is 1/2. The probabilities of a composite event and its complement always add up to 1.

   (a) Suppose your composite event E includes all aces in a well-shuffled deck. What is p(E)? _____

   (b) What is included in the complement of E? _____

– – – – – – – – – – – – – – –

(a) 4/52 or 1/13; (b) all cards except aces

19. If you know p(E), you can find the probability of the complement of E by using the formula 1 − p(E) = p(complement).
   What is the probability of the complement of the composite event "all aces in the deck"? _____

– – – – – – – – – – – – – – –

12/13 (1 − 1/13 = 12/13)

20. Suppose the composite event E consists of 1 or 2 on a toss of a fair die.

    (a) What is p(E)? _____

    (b) What is the complement of E? _____

    (c) What is p(complement)? _____

    - - - - - - - - - - - - - -

    (a) 1/3 (2/6); (b) 3, 4, 5, 6; (c) 2/3 (1 − 1/3)

21. Suppose you have a box that contains ten marbles: three white, two red, and five blue. What is the probability that you pull out:

    (a) a red marble? _____

    (b) a blue marble? _____

    - - - - - - - - - - - - - -

    (a) 1/5; (b) 1/2

22. Consider the marbles of frame 21. Suppose E is "red or blue." The complement of E:

    ___(a) has a probability of 1/5
    ___(b) has a probability of 3/10
    ___(c) is white
    ___(d) consists of all marbles that are not red or blue

    - - - - - - - - - - - - - -

    (b); (c); (d)

You now know the underlying principles of probability theory:

   • $0 \leq p(e) \leq 1$
   • The sum of all possible p(e)'s in $\mathcal{E}$ is 1.
   • p(E) = the sum of p(e)'s in E

So far we have discussed only one-part examples. Now we shall progress to more complex—and more interesting—problems.

PROBABILITY OF UNRELATED EVENTS

23. When you toss one die, the event set $\mathcal{E}$ consists of six simple events: 1, 2, 3, 4, 5, and 6. Suppose you toss two dice. The number of the event set now is 6 x 6, or 36. For each result of one die there are:

___(a) 36 different results for the other
___(b) 6 different results for the other

– – – – – – – – – – – – – –

(b)

24. The number showing on the face of one die has no effect on the number showing on the face of the other die. The two parts are:

___(a) related
___(b) unrelated
___(c) complements

– – – – – – – – – – – – – –

(b)

25. When simple events from two unrelated experiments are considered together, you multiply the numbers of the separate event sets to get the number of the total set. Thus, the event for tossing two dice is 6 x 6, or 36. The number of elements in the event set for flipping a

penny and a nickle is _____.

– – – – – – – – – – – – – –

2 x 2, or 4

26. In dice games, it is frequently the sum of the points on the two dice that determines the outcome of the game. The boxes on page 186 show the possible results when two dice are tossed. The total sum 2 occurs in only one of the thirty-six possible results: $p(2) = 1/36$. The sum 4 occurs in three of the thirty-six possible results: $(1,3)$, $(2,2)$, and $(3,1)$; $p(4) = 3/36 = 1/12$. Find the following:

(a) $p(7) =$ _____

(b) $p(9) =$ _____

(c) $p(11) =$ _____

– – – – – – – – – – – – – –

(a) $3/36 = 1/6$; (b) $4/36 = 1/9$; (c) $2/36 = 1/18$

(1,1)	(1,2)	(1,3)	(1,4)	(1,5)	(1,6)
(2,1)	(2,2)	(2,3)	(2,4)	(2,5)	(2,6)
(3,1)	(3,2)	(3,3)	(3,4)	(3,5)	(3,6)
(4,1)	(4,2)	(4,3)	(4,4)	(4,5)	(4,6)
(5,1)	(5,2)	(5,3)	(5,4)	(5,5)	(5,6)
(6,1)	(6,2)	(6,3)	(6,4)	(6,5)	(6,6)

POSSIBLE RESULTS WHEN TWO DICE ARE TOSSED

2	3	4	5	6	7
3	4	5	6	7	8
4	5	6	7	8	9
5	6	7	8	9	10
6	7	8	9	10	11
7	8	9	10	11	12

SUMS OF TOSSES WHEN TWO DICE ARE TOSSED

COMBINATIONS AND THEIR PROBABILITIES

Now we are going to consider combinations. How many different groups of two men can be taken from a population of seven? We draw a sample of size s from a population of size N. The number of different samples formed is given by $\binom{N}{s}$.

Suppose you have three books on your desk. You need to take two of them to class. You grab two without looking as you run out the door. How many different combinations might you have grabbed? What is the probability that you got the right combination?

Let us see what the combination possibilities are. Call the books A, B, and C. You might get A and B, or B and C, or A and C. Thus, there are three possible combinations. The right combination was A and B. You had one possible success out of three possible events. So p(e) = 1/3.

In this section we will concentrate on finding the number of combinations of s objects (s = 2 in the example of the books) from a population of size N (N = 3 in this example). You will learn to use a table to find the number of possible combinations.

27. Consider a box of six marbles: one each of red, white, blue, yellow, green, and orange. You want to select two at random. In order to find the probability of drawing any specific combination, you must first know how many possible combinations of two (or samples of two) can be obtained from a population of six.

 (a) The sample size s is _____ .

 (b) The population size N is _____ .

– – – – – – – – – – – – – – –

 (a) 2;   (b) 6

28. The expression $\binom{N}{s}$ is used in finding numbers of combinations. In the problem of the last frame, this is $\binom{6}{2}$. Now look at the table on page 188 to find the number of combinations. Find the specific value of N in the vertical column and the value of s in the horizontal row.

 The value of $\binom{6}{2}$ is _____ .

– – – – – – – – – – – – – –

 15 (You could select fifteen different combinations of two marbles from a group of six marbles.)

VALUES OF $\binom{N}{s}$

N \ s	1	2	3	4	5	6	7	8	9	10
1	1									
2	2	1								
3	3	3	1							
4	4	6	4	1						
5	5	10	10	5	1					
6	6	15	20	15	6	1				
7	7	21	35	35	21	7	1			
8	8	28	56	70	56	28	8	1		
9	9	36	84	126	126	84	36	9	1	
10	10	45	120	210	252	210	120	45	10	1

The numbers in the table above are symmetrical for each N if a column for s = 0 is included, since $\binom{N}{0}$ = 1. When N = 8, for example, the $\binom{N}{s}$ numbers are 1, 8, 28, 56, 70, 56, 28, 8, 1. These values can be derived from what is known as Pascal's triangle.

```
                    1
                 1     1
              1     2     1
           1     3     3     1
        1     4     6     4     1
     1     5    10    10     5     1
  1     6    15    20    15     6     1
1     7    21    35    35    21     7     1
```

Each number is the sum of the two numbers just to the right and left in the row above it: 7 is the sum of 6 and 1; 35 is the sum of 15 and 20; 1 is the sum of 1 and 0. This triangle can be continued indefinitely. A computer can be programmed to calculate the values for any N.

Why would you want to know the larger values? Do you know the number of possible bridge hands? It is $\binom{52}{13}$ = 635,013,559,600. The number of possible poker hands is $\binom{52}{5}$ = 2,598,960.

29. The fifteen combinations of two marbles that can be obtained from the box are listed below.

red, white	white, yellow	yellow, green
red, yellow	white, blue	yellow, orange
red, blue	white, green	blue, green
red, green	white, orange	blue, orange
red, orange	yellow, blue	green, orange

Which of the following statements is true?

___(a) The order of events in a combination is not important.
___(b) (Red, blue) and (blue, red) are different combinations.
___(c) More of these combinations contain red than contain blue.

– – – – – – – – – – – – – –

(a)

30. Using the table on page 188, find $\binom{N}{s}$ for each of the following.

(a) a sample of one student from a population of seven _____

(b) a sample of seven students from a population of seven _____

(c) a sample of two men from a population of nine _____

(d) a sample of four women from a population of eight _____

– – – – – – – – – – – – – –

(a) $\binom{7}{1} = 7$; (b) $\binom{7}{7} = 1$; (c) $\binom{9}{2} = 36$; (d) $\binom{8}{4} = 70$

31. $\binom{N}{1}$ is always equal to (refer to the table on page 188):

___(a) 1
___(b) s
___(c) N

– – – – – – – – – – – – –

(c)

32. If N is equal to s, $\binom{N}{s}$ is equal to:

___(a) 1
___(b) s
___(c) N

– – – – – – – – – – – – –

(a) (If N and s both equal 1, then $\binom{N}{s} = 1 = N = s$.)

33. Read the examples below. Decide the values of N and s. Then refer to the table on page 188 and write the value of (^N_s).

(a) There are ten houses on the block. You need a sample of five houses. How many different combinations of five houses could you obtain? _____

(b) You need to select three horses from a stable of four. How many combinations can you obtain? _____

(c) You have seven bills to pay, but want to pay only five this week. How many combinations do you have to select from? _____

- - - - - - - - - - - - - - -

(a) 252; (b) 4; (c) 21

34. Suppose you really do intend to pay five out of seven bills. You select the five at random. What is the probability of a specific set of five bills being paid? _____

- - - - - - - - - - - - - - -

1/21 (one event out of 21 possible sets)

35. Ten people enter a local contest. Three of them will be chosen randomly to win a trip to Las Vegas.

(a) How many combinations are possible? _____

(b) What is the probability that a specific group of three will go?

_____

(c) What is the probability that any one specific person will go?

_____

- - - - - - - - - - - - - - -

(a) 120; (b) 1/120; (c) 1/10 (Each person has an equal chance to be selected.)

PERMUTATIONS AND THEIR PROBABILITIES

36. The samples we have discussed have been unordered; now we shall consider ordered samples. (A, B) and (B, A) are the same combination of events. Combinations are unordered. (A, B) and (B, A) are two _permutations_ of A and B. Permutations are:

___(a) the same as combinations
___(b) ordered
___(c) unordered

- - - - - - - - - - - - - - -

(b)

37. Match the following.

___(a) (A, B, C) (B, C, D) (A, B, D)
___(b) (A, B, C) (B, C, A) (C, B, A)
___(c) ordered
___(d) unordered

1. combinations
2. permutations
3. probability

- - - - - - - - - - - - - -

(a) 1; (b) 2; (c) 2; (d) 1

38. The number of permutations of n different objects is equal to 1 x 2 x 3 x . . . x n, or n! (read "n factorial"). Thus, there are six (1 x 2 x 3) possible permutations of three objects. Three different pictures could be hung on a wall in six different ways:

   A B C   A C B   B C A   B A C   C B A   C A B

   Find the number of possible permutations for each of the following values of n.

(a) 5 _____

(b) 2 _____

(c) 10 _____

- - - - - - - - - - - - - -

(a) 5! = 120; (b) 2! = 2; (c) 10! = 3,628,800

39. In how many ways can four people be lined up for a photograph?

_____

- - - - - - - - - - - - -

4! = 24

40. Consider four people: A, B, C, and D. These four are to be seated along a bench. What is the probability that the arrangement will be either (A B C D) or (D C B A)? _____

1/12 (1/24 + 1/24)

41. In the example of frame 40, what is the probability that the first person on the right is A? _____

1/4 (Each person has an equal chance to be on the right.)

42. Two of our four people (A, B, C, and D) are to be selected at random for testing and then photographed again. Now we need to know how many permutations there are of N objects taken s at a time. This is symbolized $(N)_s$. In this example, N = 4 and s = 2. Two steps are involved in calculating $(N)_s$:

   Step 1: How many samples of size s can be selected from a population of size N?
   Step 2: How many permutations of the sample size s are possible?

In this problem $(N)_s = (4)_2$.

   (a) What is $\binom{N}{s}$ in the table on page 188? _____

   (b) How many permutations are possible of two objects? _____

(a) 6; (b) 2 (s! = 2! = 2)

43. $(N)_s$, the number of permutations of N objects taken s at a time, equals $\binom{N}{s}$ x s!. If $\binom{N}{s} = \binom{4}{2}$, what is the value of $(N)_s$?

_____

----

12 (6 x 2)

44. The probability of the occurrence of $\binom{N}{s}$ is $\dfrac{1}{\binom{N}{s}}$. The probability of $(N)_s$ is $\dfrac{1}{(N)_s}$.

Referring again to the example in frame 42, what is the probability that the final photo contains people A and C, in that order?

_____

– – – – – – – – – – – – – –

1/12

45. Suppose you have five different colored pens (red, green, blue, black, and brown) in your briefcase. You reach in, remove three, and hand them to three friends (A, B, and C). What is the probability that the pens you removed were red, green, and blue?

_____

– – – – – – – – – – – – – –

1/10 $[\binom{N}{s} = \binom{5}{3} = 10;\ p\binom{5}{3} = 1/10]$

46. Refer to the problem in frame 45. What is the probability that the three pens removed were red, green, and blue, and that A received red, B received green, and C received blue? Remember:

$$(N)_s = \binom{N}{s} \times s!;\quad p\,[(N)_s] = \dfrac{1}{(N)_s}$$

_____

– – – – – – – – – – – – – –

1/60 $[(N)_s = 10 \times 6;\ p(60) = 1/60]$

Probability problems in real life are often much more complex than those presented here. The numbers involved are much larger and the options may not be equally likely. We won't concern outselves with weighting the alternatives, but we will look briefly at the flowcharting of probability problems.

47. N = 9; s = 3. Match the following.

___(a) $\binom{N}{s}$ 1. number of possible ordered samples

___(b) $(N)_s$ 2. number of possible unordered samples
          3. number of permutations of N objects
___(c) $(N)_N$ 4. = N!

– – – – – – – – – – – – –

(a) 2; (b) 1; (c) 3, 4

48. In frame 43 we used one formula for calculating $(N)_s$: $(N)_s = \binom{N}{s} \times s!$. An equivalent formula for $(N)_s$, which does not require that you look up $\binom{N}{s}$ on a table, is:

$$(N)_s = N \times (N - 1) \times (N - 2) \times \ldots \times (N - s + 1)$$

For example, consider $(9)_4$. In this case, the last term is $(9 - 4 + 1) = 6$. The expression for $(N)_s$ is $(9)_4 = 9 \times 8 \times 7 \times 6$, and the value of $(9)_4$ is 3024.

Which of the expressions below would you use to calculate $(13)_3$?

____(a) $(13)_3 = 13 \times 12 \times 11 \times 10 \times 9 \times 8 \times 7 \times 6 \times 5 \times 4 \times 3$
____(b) $(13)_3 = 13 \times 12 \times 11 \times 10 \times 9$
____(c) $(13)_3 = 13 \times 12 \times 11 \times 10$
____(d) $(13)_3 = 13 \times 12 \times 11$

- - - - - - - - - - - - - - -

(d) $[(N - s + 1) = 13 - 3 + 1 = 11]$

49. If $(N)_s = N \times (N - 1) \times \ldots \times (N - s + 1)$, write the expression for $(N)_s$ in each of the following.

(a) $(8)_3 =$ _____

(b) $(4)_2 =$ _____

(c) $(14)_6 =$ _____

(d) $(215)_4 =$ _____

- - - - - - - - - - - - - - -

(a) $8 \times 7 \times 6$; (b) 4×3; (c) $14 \times 13 \times 12 \times 11 \times 10 \times 9$;
(d) $215 \times 214 \times 213 \times 212$

50. Suppose you want to know the probability of any one specific permutation of eight objects taken three at a time. First you calculate $(N)_s$. Then, $p[(N)_s] = \dfrac{1}{(N)_s}$. Find $p[(8)_3]$. _____

- - - - - - - - - - - - - - -

1/336 $[(8)_3 = 8 \times 7 \times 6 = 336]$

51. Study the flowchart below.

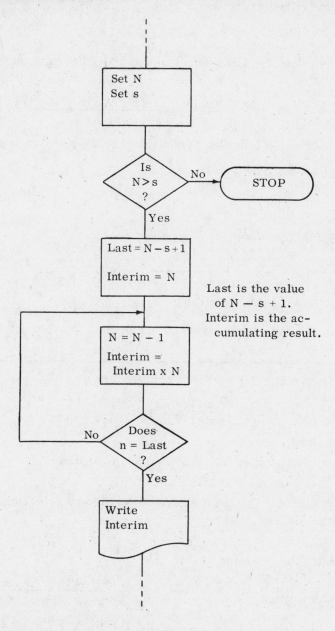

Last is the value
of N − s + 1.
Interim is the ac-
cumulating result.

This flowchart can be used to calculate:

____(a) any value of $(N)_s$

____(b) $(N)_s$ when N is larger than s

____(c) $\binom{N}{s}$ if N > s

– – – – – – – – – – – – –

(b)

52. The value of $\binom{N}{s}$ can be calculated without using a table. The formula is:

$$\binom{N}{s} = \frac{N \times (N-1) \times \ldots \times (N-s+1)}{1 \times 2 \times 3 \ldots \times s}$$

Which of the following is equivalent to the above formula?

____(a) $\binom{N}{s} = (N)_s \times s!$

____(b) $\binom{N}{s} = \frac{N!}{s!}$

____(c) $\binom{N}{s} = \frac{(N)_s}{s!}$

– – – – – – – – – – – – –

(c)

53. Use the formula in frame 52 to calculate the value of $\binom{12}{2}$.

_____

– – – – – – – – – – – – –

66 $[\binom{N}{s} = \binom{12}{2} = \frac{\cancel{12}^{6} \times 11}{1 \times \cancel{2}_{1}} = 66]$

54. Use your answer in frame 50 to help calculate $\binom{8}{3}$. _____

– – – – – – – – – – – – –

56 (336/6. This answer can be verified by referring to the table on
page 188.)

55. Match the following.

____(a) $(N)_s$ 1. $\dfrac{N \times (N-1) \times \ldots \times (N-s+1)}{1 \times 2 \times \ldots \times s}$

____(b) $\binom{N}{s}$ 2. $N \times (N-1) \times \ldots \times (N-s+1)$

          3. $N! \times s!$

– – – – – – – – – – – – –

(a) 2; (b) 1

56. $\binom{N}{s}$ can also be calculated using a flowchart, as shown on page 198.

> LAST is the last term in the numerator $(N - s + 1)$.
> INTERIM is a partial product.
> SFACT is s! (s factorial).

(a) Which section of the flowchart on page 198 calculates s! ?

(A, B, or C) _____

(b) Which section calculates $(N)_s$? _____

(c) Which section calculates $\dfrac{(N)_s}{s!}$? _____

- - - - - - - - - - - - - -

(a) B; (b) A; (c) C

57. Calculate the following values of $\binom{N}{s}$.

(a) $\binom{13}{4}$ _____

(b) $\binom{14}{6}$ _____

(c) $\binom{4}{2}$ _____

- - - - - - - - - - - - - -

(a) 715 $\left(\dfrac{13 \times \cancel{12}^{\,\cancel{}} \times 11 \times \cancel{10}^{\,5}}{\cancel{2} \times \cancel{3} \times \cancel{4}} = 13 \times 11 \times 5\right)$

(b) 3003 $\left(\dfrac{\cancel{14}^{\,7} \times 13 \times \cancel{12}^{\,2} \times 11 \times \cancel{10}^{\,2} \times \cancel{9}^{\,3}}{1 \times \cancel{2} \times \cancel{3} \times \cancel{4} \times \cancel{5} \times \cancel{6}} = 7 \times 13 \times 11 \times 3\right)$

(c) 6 $\left(\dfrac{\cancel{4}^{\,2} \times 3}{\cancel{2}} = 6\right)$

58. Find the probabilities of:

(a) any specific unordered sample of four objects from a population of thirteen _____

(b) any specific ordered sample of three objects from a population of fourteen _____

- - - - - - - - - - - - - -

(a) 1/715; (b) 1/2184 $[(14)_3 = 14 \times 13 \times 12]$

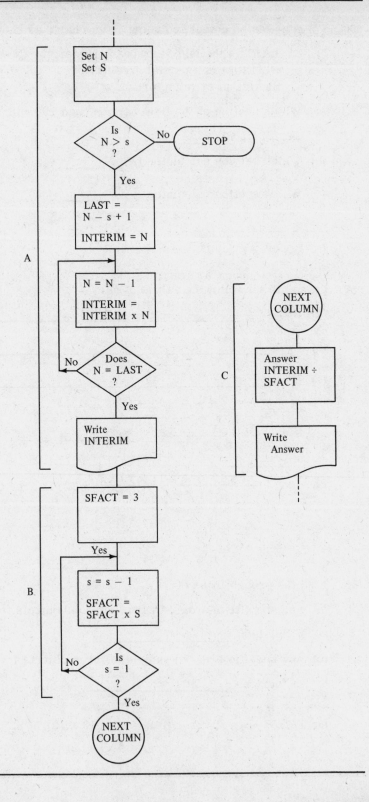

In this chapter you learned to calculate the probabilities of simple unrelated events. You then learned to differentiate between combinations and permutations and to calculate the probability of any specific combination or permutation, given sample size and population size.

Only equally likely cases have been considered in this chapter. The concepts remain the same even when the events are not equally likely.

Random numbers are often used in connection with probability. Few people are called upon to generate random numbers since effective programs already carry out this function. If you do not wish to pursue the following section, which discusses the basic characteristics of random numbers, turn directly to the Self-Test on page 206.

RANDOM NUMBERS (Optional Section)

The laws of mathematical probability are applicable if and only if a sample from a population is obtained at random. In the problems we discussed earlier in this chapter, we specified random samples by saying a well-shuffled deck, a fair die, and mixed-up marbles. These game devices can be adapted for use in more practical problems in order to achieve a random sample. For example, to select one of six men to go to a convention, you could assign each man a number from 1 to 6, then roll a die.

In this section, we will discuss random numbers, their uses, and how they are generated.

59. The laws of probability apply only when a sample is random. The randomness refers to the way in which the sample is selected. Each individual must have an equal chance of being selected.

Which of the following ways of selecting a sample might produce a random sample of undergraduates in a state university?

____(a) Select every tenth freshman.
____(b) Select all math or psychology undergraduates.
____(c) Select all undergraduates whose last names begin with any of four specific letters.

— — — — — — — — — — — — — — —

(c)

60. When samples must be taken from a relatively large population, a number may be assigned to each member. The quantity of numbers required is then selected randomly. Which of the following ways of selecting a number is random?

___(a) Write each number on a separate slip of paper, put the papers in a bowl, shake them up, and draw out a number.

___(b) Ask someone to look over the list of numbers and select one.

_ _ _ _ _ _ _ _ _ _ _ _ _ _

(a)

61. The term random in reference to a sequence means that no set pattern should be apparent. Which of the sequences below has no apparent pattern?

___(a) 1, 2, 3, 4, 5
___(b) 2, 4, 6, 8, 10
___(c) 1, 7, 4, 3, 8, 2

_ _ _ _ _ _ _ _ _ _ _ _ _ _

(c)

62. Each number in a random sequence of digits should be equally likely to appear. Which of the sequences below might be random for the digits 0 to 9?

___(a) 1, 1, 1, 1, 2, 2
___(b) 4, 3, 1, 7, 9, 8, 2, 5, 0, 6
___(c) 3, 7, 1, 9, 5, 1

_ _ _ _ _ _ _ _ _ _ _ _ _ _

(b) (Choice (c) contains only odd numbers.)

63. In order for a sequence to be random each number must be

_____ _____ to appear, and there must

be no _____ _____.

_ _ _ _ _ _ _ _ _ _ _ _ _ _

equally likely; set pattern (or apparent order)

The selection of a number at random is not a trivial problem. For example, studies have shown that when a person is asked to name a number between 1 and 10, he is more likely to say 3 or 7 than 4 or 8. A ten-sided die, with digits from 0 to 9, has been developed as a random-digit generator. When the die is thrown, there is equal probability that each face will come up. This device can be used to manually generate a small number of random digits. Next we will consider how to get a computer to generate random numbers.

Some statistical methods require great quantities of random numbers. Computers can generate "pseudo-random" numbers that behave like random numbers. This is adequate for most purposes. The numbers are pseudo-random rather than random because the list can be regenerated in the same order. These numbers appear to be selected randomly.

64. The flowchart on page 202 can be used to generate the numbers from 1 to 10 in pseudo-random order. The value assigned to x in the beginning must be an integer value between 1 and 10. Assume x = 7. Follow the flowchart and write the next three numbers in the sequence.

(a) n(1) = 7

(b) n(2) = _____

(c) n(3) = _____

(d) n(4) = _____

- - - - - - - - - - - - - - -

(a) 7; (b) 3; (c) 6; (d) 1 (These are pseudo-random since the same list is guaranteed every time the algorithm is used.)

65. In the flowchart on page 202, the use of the number 11 gives us ten random numbers. Then the cycle repeats itself in the same order. It is desirable, then, to get a longer cycle length since a pattern shows up with repetition. Consider the flowchart using 13 instead of 11. This will create a pseudo-random sequence of the numbers from 1 to 12. If you start with x = 7 again, the third number will

be _____.

- - - - - - - - - - - - - - -

2 (the second is 1)

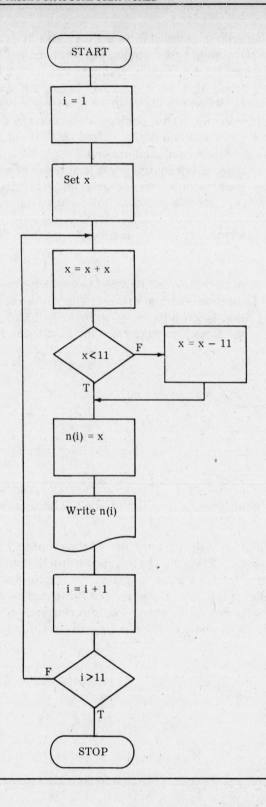

66. The cycle length of a pseudo-random sequence is:

___(a) the number of items before the sequence repeats itself
___(b) ten
___(c) the number of times a specific sequence is repeated

– – – – – – – – – – – – – – –

(a)

67. The method used in frame 64 works only with certain numbers. It worked with 11 and 13, both of which are prime numbers. (A prime number can be evenly divided only by itself and 1.) It works, however, only with certain prime numbers. The number 17 does not produce a pseudo-random sequence. Trace through the flowchart on page 202 with the number 17, starting with 7. What is its cycle

length? _____

– – – – – – – – – – – – – –

8 (7, 14, 11, 5, 10, 3, 6, 12. If you start with a number not in this sequence, you get the other eight integers between 1 and 16.)

68. The sequence generated in the previous frame is not pseudo-random because all of the numbers from 1 to 16 are

_____.

– – – – – – – – – – – – – –

not equally likely (You needn't be too concerned about how to determine which prime numbers can be used to generate pseudo-random sequences. The critical thing is that some primes do work—such as 11, 13, 19, 29, 37, 53, and 59. Some larger workable primes are 941 and 947. We shall use these later.)

69. The pseudo-random sequences generated by 11 and 13 are given below:

    11: 8, 5, 10, 9, 7, 3, 6, 1, 2, 4
    13: 8, 3, 6, 12, 11, 9, 5, 10, 7, 1, 2, 4

Another way of generating a pseudo-random sequence is to calculate two sequences simultaneously, add the items individually, and use the sums as the random sequence. The cycle length will be 60 if you do this with the sequences above of ten and twelve digits each.

(a) The largest number in the new sequence will be _____.

(b) The first four numbers are _____, _____, _____, and _____.

_ _ _ _ _ _ _ _ _ _ _ _ _ _

(a) 22 (The highest sum is 10 + 12.); (b) 16, 8, 16, 21

70. The flowchart on page 205 shows the procedure for writing a random-number generating program. Study the flowchart.

(a) How many numbers will be generated? _____

(b) What is the largest possible number? _____

(c) What is the smallest possible number? _____

(d) Write the first two numbers: N(1) = _____; N(2) = _____

_ _ _ _ _ _ _ _ _ _ _ _ _ _

(a) 5000; (b) 999; (c) 2 (1 + 1); (d) N(1) = 801; N(2) = 602

A table of primes that can be used in generating pseudo-random numbers using this method is found in A Second Course in Number Theory by Harvey Cohn (New York: John Wiley & Sons, 1962). More detailed explanations of the method can be found in Mathematics and Computing with FORTRAN Programming by William Dorn and Herbert Greenberg (New York: John Wiley & Sons, 1967).

Now take the Self-Test on page 206. (The material covered in this optional section is not included in the Self-Test.)

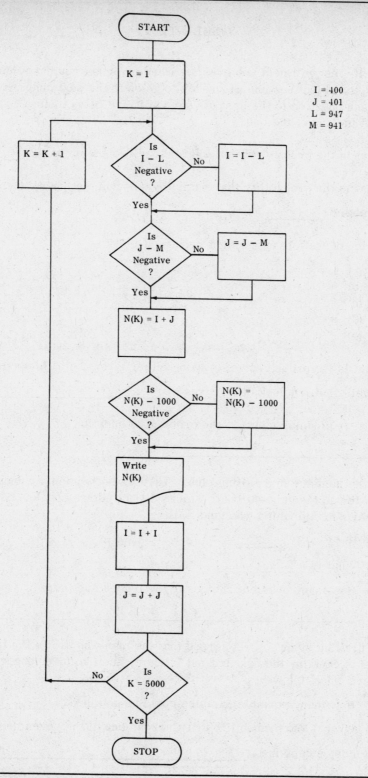

SELF-TEST

This Self-Test will help you evaluate whether or not you have mastered the chapter objectives and are ready to go on to the next chapter. Answer each question to the best of your ability. Correct answers are given at the end of the test.

1. What is the probability that a fair die will come up on 6? _____

2. What is the probability that a fair die will come up on an even

    number? _____

3. Which of the following might represent p(e)?

    ___(a) 1.5
    ___(b) 0
    ___(c) 1
    ___(d) $-\frac{1}{2}$
    ___(e) $\frac{1}{4}$

4. What is the probability of drawing either a heart or a black ace from

    a well-shuffled deck of 52 cards? _____

5. What is p(complement) in the problem of question 4?

    _____

6. Seven men are in a sailing club. They all show up on a Sunday morning to use the only sailboat, which will hold three people. They draw straws to determine who goes sailing.

    (a) What is N? _____

    (b) What is s? _____

    (c) How many possible choices for who goes are there?

    _____

7. In drawing straws, it is agreed that the man who draws the longest one is captain, the next longest is mate, and the third longest just stays out of the way.

    (a) How many possible samples could be obtained? _____

    (b) What is the probability that Carson, one of the seven men, be-

        comes captain? _____

8. Calculate the following.

(a) $p\binom{9}{2}$ = _____

(b) $p\left[(9)_2\right]$ = _____

(c) $p\binom{17}{3}$ = _____

(d) $p\left[(4)_4\right]$ = _____

ANSWERS TO SELF-TEST

Compare your answers to the Self-Test with the correct answers given below. If all of your answers are correct, you are ready to go on to the next chapter. If you missed any questions, study the frames indicated in parentheses following the answer.

1. 1/6   (frames 5-9)

2. 1/2 (1/6 + 1/6 + 1/6 = 3/6)   (frames 9-13)

3. (b); (c); (e) $[0 \leq p(e) \leq 1]$ (frames 5-6)

4. 15/52 (13/52 + 2/52)   (frames 9-13)

5. 37/52 [p(complement) = 1 − p(e)]   (frames 16-19)

6. (a) N = 7

    (b) s = 3

    (c) $\binom{N}{s} = \binom{7}{3} = 35$ $(\dfrac{7 \times 6 \times 5}{1 \times 2 \times 3} = 35)$ (frames 27-35, 52-56)

7. (a) $(N)_s = (7)_3 = 210$ (7 x 6 x 5)

    (b) 1/7 (7 equally likely results   (frames 48-51)

8. (a) 1/36

    (b) 1/72

    (c) 1/680

    (d) 1/24   (frames 48-58)

CHAPTER NINE

Statistics

Much of computer output is either composed of statistics or used to compile statistics. The language and concepts of statistics are indispensable in the computer world. Masses of data are briefly explained by using statistics; you must be able to comprehend these explanations.

When you complete this chapter, you will be able to:

- calculate any of the three measures of central tendency;

- state the range of a distribution;

- calculate the average deviation of a set of values from the mean;

- calculate the variance and standard deviation for a distribution;

- determine how much of a distribution is within one standard deviation from the mean.

Additional information about statistics can be found in either of the following Self-Teaching Guides:

Koosis, Donald J., Business Statistics (New York: John Wiley & Sons, 1972)

Koosis, Donald J., Statistics (New York: John Wiley & Sons, 1972)

MEASURES OF CENTRAL TENDENCY

One of the primary purposes of statistics is to condense and channel large amounts of data into meaningful figures, and to make predictions from these figures. In this section you will learn some ways of condensing masses of information into meaningful statistics.

1. The arithmetic average, or <u>mean</u>, is a statistical concept, a measure of the central tendency of a distribution. To find the mean, add up a list of numerical items, then divide by the total number of items.

    Which of the following represents a mean?

    ___(a) $\dfrac{6 + 7 + 4}{4}$

    ___(b) $1 \times 2 \times 3 \times 4$

    ___(c) $\dfrac{4 + 5 + 6}{3}$

- - - - - - - - - - - -

(c)

2. Suppose a student has taken six tests during a semester. His scores were 70, 90, 95, 80, 90, and 85. What is mean (average) score?

- - - - - - - - - - - -

85 $(\dfrac{510}{6})$

3. Sometimes you want to know the frequency distribution—how frequently each result occurred—rather than the mean. The frequency distribution for the scores in the preceding frame is shown below:

Score	Frequency
70	1
80	1
85	1
90	2
95	1

Construct a frequency distribution for the scores given below.

80, 95, 90, 70, 85, 90, 85, 80, 85

- - - - - - - - - - - -

Score	Frequency
70	1
80	2
85	3
90	2
95	1

4. What is the mean for the scores you worked with in frame 3?

_ _ _ _ _ _ _ _ _ _ _ _ _ _ _ _

84.44

5. The frequency distribution of a set of data gives you another measure, the <u>mode</u>. The mode is the item that occurs with the greatest frequency. The ages of members of a college composition class, for example, may be under consideration. The frequency distribution is given below.

Age	Frequency
17	2
18	17
19	20
20	12
21	8
22	10
23	4
27	1
33	1

(a) What is the mode? _____

(b) The total of all ages is 1500 years. What is the mean? _____

_ _ _ _ _ _ _ _ _ _ _ _ _ _ _

(a) 19 (the age that occurs with the greatest frequency)
(b) 20 (1500 divided by the sum of the frequencies)

6. The <u>median</u> is still another measure of the central tendency of a distribution. The median represents the middle; half the items are larger than the median and half are smaller.
    Find the median for the frequency distribution in frame 5.

(a) How many items are listed (sum the frequency column)? _____

(b) Half your answer to (a): _____

(c) What age is in the middle? _____

_ _ _ _ _ _ _ _ _ _ _ _ _ _

(a) 75; (b) $37\frac{1}{2}$; (c) 19 (If a median falls between two values, it is considered to be halfway between them.)

7. Suppose a true-false test with ten items is given to a class. The scores obtained by the students are as follows:

$$10, 8, 10, 10, 9, 7, 8, 7, 6, 10, 9$$

(a) Construct a frequency distribution.

(b) What is the mode? _____

(c) What is the median? _____

(d) What is the mean (to three decimal places)? _____

- - - - - - - - - - - - - -

(a)

Score	Frequency
10	4
9	2
8	2
7	2
6	1

(b) 10; (c) 9; (d) 8.545

8. The range of a distribution is a very rough measure of its variability. It is simply the difference between the highest and the lowest items in the distribution. For the distribution in the last frame, the range is _____.

- - - - - - - - - - - - - -

4 (10 − 6)

9. Here are the scores one student received on a series of eleven quizzes:

$$90, 73, 86, 82, 78, 91, 84, 72, 74, 87, 90$$

Find the following:

(a) range _____

(b) mean _____

(c) mode _____

(d) median _____

– – – – – – – – – – – – – – –

(a) 19; (b) 82.5; (c) 90; (d) 84

10. We can use algorithms and flowcharts to compute statistical measures.

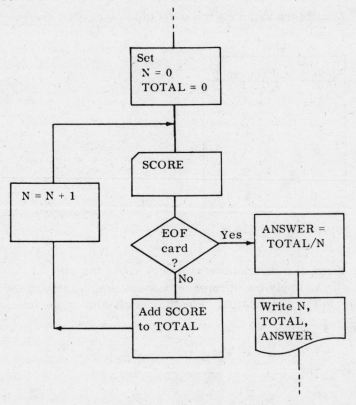

The flowchart above could be used to calculate a:

___(a) range   ___(c) median
___(b) mode   ___(d) mean

- - - - - - - - - - - - - - -

(d)

11. Suppose you have a stack of about 4000 cards, each with a test score punched into it. Each score can be any whole number from 0 to 100. Which of the following suggestions might you use to find the mode?

____(a) Add the scores while counting the cards, then divide the total by the number of cards.

____(b) Count the number of times each score occurs and use the score with the highest frequency as the mode.

____(c) Set up a frequency distribution table and determine the score that is in the middle of the distribution.

- - - - - - - - - - - - - - -

(b)

12. The median, mean, and mode can be very close together in any distribution, or they can be widely spread. Suppose the salaries of seven people are $15,000, $8,000, $48,000, $3,200, $12,000, $6,000, and $3,200. Calculate the following to the nearest dollar:

(a) mean _____

(b) median _____

(c) mode _____

(d) range _____

- - - - - - - - - - - - - -

(a) $13,629 ($\frac{95,400}{7}$); (b) $8,000; (c) $3,200; (d) $44,800

AVERAGE DEVIATION

You have now learned to calculate the three measures of central tendency—the mean, the median, and the mode. You have also learned to calculate the range of a distribution, which is a measure of its variability. In this section you will learn some better ways of stating how a set of data is distributed. To do this you must use the absolute value of a number. This is done by ignoring any positive or negative sign associated with the number.

13. The average deviation of items from the mean is a measure of variability of a distribution. This average (or mean) deviation is calculated as follows:

> Step 1: Calculate the mean of the distribution.
> Step 2: Find the amount each value varies from the mean.
> Step 3: Average the absolute value (ignoring signs) of the deviation.

Follow these three steps to find the average deviation from the mean for the data below.

$$1, \ 2, \ 2, \ 4, \ 4, \ 5$$

(a) mean _____

(b) deviation of each from mean _____ _____ _____ _____ _____

_____

(c) average deviation _____

- - - - - - - - - - - - - - - -

(a) 3 (18/6); (b) 2, 1, 1, 1, 1, 2; (c) 1.33 (8/6)

14. The process of calculating the average deviation from the mean can be written using symbols:

$$\frac{\sum\limits_{i=1}^{n} |x_i - \overline{x}|}{n}$$

An explanation is needed, however. $\overline{x}$ refers to the mean of the distribution. x_i refers to a particular bit of data. The vertical lines ($|\ |$) indicate absolute value. The Greek letter sigma ($\sum$) is used to mean the summation, or adding up. The notation below and above $\sum$ (i = 1 below and n above) explains what particular values i can take. x_i starts with the first value in the list, and continues to the nth value.

(a) In the problem of frame 13, the value of n is _____.

(b) The value of $\overline{x}$ is _____.

(c) The value of $|x_i - \overline{x}|$ is _____.

- - - - - - - - - - - - - - -

(a) 6; (b) 3; (c) 2 (not –2; $|1 - 3|$ = +2. If you do not take the absolute values in calculating this sum, the average deviation will always be zero.)

15. The formula below would be used to calculate:

$$\frac{\displaystyle\sum_{i=1}^{n} |x_i - \overline{x}|}{n}$$

___(a) the mean of a distribution
___(b) the mean deviation from a median
___(c) the variability of a distribution
___(d) a measure of central tendency

- - - - - - - - - - - - - - -

(c) (The average deviation from the mean is a measure of variability.)

16. Consider the formula and the set of scores below.

$$\frac{\displaystyle\sum_{i=1}^{n} |x_i - \overline{x}|}{n} \qquad\qquad 10,\ 7,\ 4,\ 9,\ 8,\ 9,\ 6,\ 9,\ 6,\ 2$$

Rewrite the formula, using appropriate values for $\overline{x}$ (mean) and n
(number of scores).

- - - - - - - - - - - - -

$$\frac{\displaystyle\sum_{i=1}^{10} |x_i - 7|}{10}$$

17. The first step in calculating the average deviation from the mean is
finding the mean. The second step is finding each item's (x_i) deviation from that mean. Complete the list of values of deviations of
x_i below, as i goes from 1 to 10. (Use the scores in frame 16.) Ig-
nore the signs; it is unimportant whether the score is larger or
smaller than the mean.

x_1	x_2	x_3	x_4	x_5	x_6	x_7	x_8	x_9	x_{10}
3	0	___	___	___	___	___	___	___	___

- - - - - - - - - - - - - -

3, 0, 3, 2, 1, 2, 1, 2, 1, 5

18. The third step is to find the average deviation: sum the deviations, then divide by n. The average deviation here is _____.

- - - - - - - - - - - - - - -

2 (20/10)

VARIANCE OF A DISTRIBUTION

The usual method of indicating the variability of a distribution is the standard deviation. Since the standard deviation depends on the variance, however, we shall consider the variance first.

19. The <u>variance</u> of a distribution is the average <u>squared</u> deviation from the mean. Which of the formulas below might be used to calculate the variance of a distribution?

___(a)
$$V = \left(\frac{\sum\limits_{i=1}^{n} |x_i - \bar{x}|}{n} \right)^2$$

___(b)
$$V = \frac{\sum\limits_{i=1}^{n} (x_i - \bar{x})^2}{n}$$

___(c)
$$V = \sum\limits_{i=1}^{n} (x_i - \bar{x})^2$$

- - - - - - - - - - - - - - -

(b) (The reason for using the squared deviation is beyond the scope of this book.)

20. Let us consider the variance of the problem in frames 16 through 18. You already calculated the deviation from the mean in frame 17. Now list the squared deviations below.

9, 0, 9, _____, _____, _____, _____, _____, _____, _____

- - - - - - - - - - - - - - -

9, 0, 9, 4, 1, 4, 1, 4, 1, 25

21. Now sum the squared deviations and find the average to get the variance. _____

- - - - - - - - - - - - - - - -

5.8 (58/10)

STANDARD DEVIATION

22. The standard deviation (abbreviated S. D.) is the square root of the variance. Which formula below might be used to calculate the standard deviation of a distribution?

___(a)

$$S.\ D.\ =\ \sqrt{\frac{\sum\limits_{i=1}^{n} \left|x_i - \overline{x}\right|}{n}}$$

___(b)

$$S.\ D.\ =\ \sqrt{\frac{\sum\limits_{i=1}^{n} x_i - \overline{x}}{n}}$$

___(c)

$$S.\ D.\ =\ \sqrt{\frac{\sum\limits_{i=1}^{n} (x_i - \overline{x})^2}{n}}$$

- - - - - - - - - - - - - - -

(c)

23. Match the following.

___(a) variance

___(b) average deviation from mean

___(c) standard deviation

1. absolute values of deviations are summed
2. deviations are squared, then summed
3. square root of mean is taken
4. measures variability

- - - - - - - - - - - - - -

(a) 2, 4; (b) 1, 4; (c) 2, 3, 4

24. Suppose six babies are born in a hospital during one shift. Their lengths are 18, 19, 21, 21, 21, and 26 inches. Use the formulas below as necessary to find the following measures.

$\dfrac{\sum\limits_{i=1}^{n} \lvert x_i - \bar{x} \rvert}{n}$	$\dfrac{\sum\limits_{i=1}^{n} (x_i - \bar{x})^2}{n}$	$\sqrt{\dfrac{\sum\limits_{i=1}^{n} (x_i - \bar{x})^2}{n}}$
Average Deviation	Variance	Standard Deviation

(a) average deviation _____

(b) variance _____

(c) standard deviation (approximate) _____

- - - - - - - - - - - - - -

(Note: Mean length is 21 inches.)
(a) 1.67 (10/6); (b) 6.3 (38/6); (c) 2.5 ($\sqrt{6.3}$ If you don't remember how to find a square root, don't be concerned. In practice you would use a table of square roots, found in most statistics books.)

25. The standard deviation of 2.5 tells us that, except for one baby, all were within one standard deviation of the mean for the distribution of baby lengths. The other baby was within how many standard deviations? _____

- - - - - - - - - - - - - -

2 (5 inches)

26. The figure on page 221 shows the relationship of standard deviation to what is called a normal curve. Which of the following statements is true?

____(a) All values in a normal curve are within three standard deviations from the mean.
____(b) More than 90 percent of the values in a normal curve are within three standard deviations from the mean.
____(c) 95.4 percent of the values in a normal curve are within one standard deviation of the mean.

- - - - - - - - - - - - - -

(b)

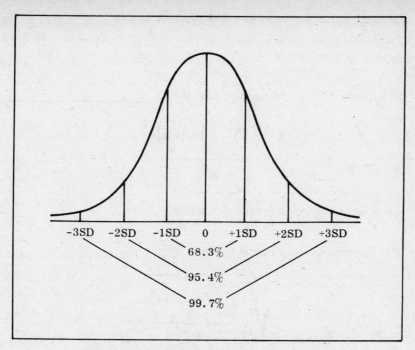

-3SD   -2SD   -1SD   0   +1SD   +2SD   +3SD

68.3%

95.4%

99.7%

Standard Deviation and a Normal Curve

The normal distribution curve represents the mean of a large number of separate samples. Many variables, including human intelligence, height, scores on tests, reaction times, and so on, show a normal distribution over a large population. In intelligence, for example, approximately 4.6 percent of the population is not within two standard deviations of the mean. Around 2.3 percent is at each extreme of the intelligence scale.

For further information on the uses and characteristics of the normal curve and standard deviation, consult the books mentioned on page 209.

REVIEW FRAMES

27. The value that occurs most often in a frequency distribution is called

the _____.

- - - - - - - - - - - - - - -

mode

28. The value that is at the halfway point in a frequency distribution is

the _____.

- - - - - - - - - - - - - -

median

29. The arithmetic average of a distribution is the _____.

- - - - - - - - - - - - - -

mean

30. The formula below is used to find the _____.

$$\frac{\sum\limits_{i=1}^{n} |x_i - \overline{x}|}{n}$$

- - - - - - - - - - - - -

mean deviation (or average deviation)

31. The formula below is used to find the _____.

$$\sqrt{\frac{\sum\limits_{i=1}^{n} (x_i - \overline{x})^2}{n}}$$

- - - - - - - - - - - - -

standard deviation

32. The range of a distribution is the _____
_____.

- - - - - - - - - - - - -

difference between highest and lowest values

33. Write the formula for calculating variance.

- - - - - - - - - - - - -

$$\frac{\sum\limits_{i=1}^{n} (x_i - \overline{x})^2}{n}$$

You have now learned to calculate the common measures of central tendency, as well as the usual measures of variability of a distribution. There is, of course, much more to statistics than this. But you have learned the terminology and notation to enable you to function in basic statistics work.

SELF-TEST

This Self-Test will help you evaluate whether or not you have mastered the chapter objectives and are ready to go on to the next chapter. Answer each question to the best of your ability. Correct answers are given at the end of the test.

The values below represent scores of fifteen students on a quiz with twenty items.

17, 19, 20, 20, 20, 16, 18, 19, 14, 17, 20, 6, 18, 16, 15

1. What is the mode? _____

2. What is the median? _____

3. What is the mean? _____

4. What is the range? _____

5. What is the average deviation from the mean? _____

6. What is the variance? _____

7. What is the (approximate) standard deviation? _____

8. How many of the scores are within one standard deviation of the mean? _____

ANSWERS TO SELF-TEST

Compare your answers to the Self-Test with the correct answers given below. If all of your answers are correct, you are ready to go on to the next chapter. If you missed any questions, study the frames indicated in parentheses following the answer.

1. 20 (frames 4-5)

2. 18 (frames 6-7)

3. 17 (255/15) (frames 1-4)

4. 14 (20 − 6) (frames 8-9)

5. 2.4 (36/15) (frames 13-28)

6. 12.1 (182/15) (frames 19-21)

7. 3.5 ($\sqrt{12.1}$) (frames 22-25)

8. 14 (only the score "6" is further away) (frame 25)

CHAPTER TEN
Linear Equations

Many of the problems facing programmers and decision-makers are concerned with discovering something that is unknown. When given certain types and amounts of information, one is expected to find out how much one should stock, or what the optimal production rate is. A simple linear equation is used to solve some problems, a system of several related equations to solve others. This chapter is primarily concerned with the basic math involved in solving simple systems of linear equations. The next chapter shows how linear equations are applied in business and science—and how they are dealt with in the computer world.

When you complete this chapter, you will be able to:

- identify linear equations in one or more variables;

- solve linear equations in one variable;

- solve linear equations in two variables using a method of substitution;

- solve linear equations in two variables using a method of comparison.

LINEAR EQUATIONS IN ONE VARIABLE

Linear equations are called linear because when drawn on a graph they are always represented by a line. This book is not concerned with graphing, but linear equations have other things in common besides their graphs, as you shall discover.

1. A linear equation is always a "first degree" equation; the terms are always raised to the first power. (Terms are never squared, cubed, etc.) Which of the following might be a linear equation?

    ___(a) $3x^2 + 9x + 8 = 0$
    ___(b) $4x - 7 = 0$
    ___(c) $4x + 3y = 9$

   - - - - - - - - - - - - - -

   (b); (c) (Equation (a) contains a squared term, thus it is not linear.)

2. A linear equation can contain any number of unknown quantities. In the linear equation $4x - 7 = 0$, there is one item that varies; this is the variable x. In $4x + 3y = 9$, there are two variables, x and y.
    Write the number of variables in each of the following linear equations.

    ___(a) $9x + 2y + 7 = 0$
    ___(b) $17x - 12 = 0$
    ___(c) $6x = 5$
    ___(d) $8x - 13y + 2z = 7$

   - - - - - - - - - - - - - -

   (a) 2; (b) 1; (c) 1; (d) 3

3. The solution to a linear equation in one variable can usually be obtained by simple algebra. Rewrite the equation so the term with the variable is on one side of the equal sign and the constant value is on the other. Then divide both sides by the coefficient of the variable. For example:

$$17x - 12 = 0$$

rewrite $\longrightarrow$ $\qquad 17x = 12$

divide both sides by 17 $\rightarrow$ $\qquad \dfrac{17x}{17} = \dfrac{12}{17}$

answer $\qquad\qquad\qquad x = \dfrac{12}{17}$

Use the two-step method just illustrated to find the value of x in the following linear equations.

(a) $4x - 4 = 0$ rewrite ⟶ _____

divide by 4 ⟶ _____

(b) $9x + 6 = 0$ rewrite ⟶ _____

divide by 9 ⟶ _____

- - - - - - - - - - - - -

(a) $4x = 4$

$x = 1$ $(\dfrac{4x}{4} = \dfrac{4}{4})$

(b) $9x = -6$

$x = -\dfrac{2}{3}$ $(\dfrac{9x}{9} = \dfrac{-6}{9})$

4. Solve the linear equations in one variable given below.

(a) $8x - 9 = 3$

(b) $2x + 1 = x - 7$

(c) $12y - 8 = 3y$

- - - - - - - - - - - - -

(a) $x = 1\dfrac{1}{2}$ $(8x = 12)$; (b) $x = -8$ $(2x - x = -7 - 1)$;

(c) $y = \dfrac{8}{9}$ $(12y - 3y = 8;\ 9y = 8)$

5. A linear equation containing one variable can always be written in the form $ax + b = 0$ when $a \neq 0$. In this form, a is the coefficient (multiplier) of the variable, x is the variable, and b is a constant value. For example, $8x - 9 = 3$ can be written $8x + (-12) = 0$, or $8x - 12 = 0$.

(a) What is the value of a in the example above? _____

(b) What is the value of b? _____

(c) What is the value of x? _____

— — — — — — — — — — — — —

(a) 8; (b) –12; (c) $1\frac{1}{2}$

6. Write the linear equations given below in the standard form (ax + b = 0).

(a) 14x − 8 = 3x + 1 _____

(b) 6 = 17x _____

(c) 12x = 4x + 7 _____

— — — — — — — — — — — — —

(a) 11x − 9 = 0 (14x − 3x − 8 − 1 = 0)
(b) 17x − 6 = 0 (or –17x + 6 = 0)
(c) 8x − 7 = 0 (12x − 4x − 7 = 0)

7. When linear equations are written in the standard form, the value of the variable can be found directly.

$$ax + b = 0$$

$$ax = -b$$

$$x = \frac{-b}{a}$$

Since $\underline{a} \neq 0$, this can always be calculated. Use the formula $x = \dfrac{-b}{a}$ to find the value of x for each of the following equations.

(a) 11x − 9 = 0 _____

(b) 17x − 6 = 0 _____

(c) 4x = −12 _____

— — — — — — — — — — — — —

(a) $\dfrac{9}{11}$ $(-\dfrac{-9}{11})$; (b) $\dfrac{6}{17}$ $(-\dfrac{-6}{17})$; (c) −3 $(-\dfrac{12}{4})$

8. Suppose you have an old-fashioned plumber who works for $7 an hour plus a basic $10 per call. He comes to your home, does the required work, and presents you with a bill for $38. How many hours did he work? You can set up a linear equation to find the answer. $7x + 10$ can represent his fee, where x is the number of hours worked. $7x + 10 = 38$ is a linear equation in one variable.

(a) Rewrite the equation in standard form: _____

(b) Since $x = \dfrac{-b}{a}$, $x =$ _____ .

- - - - - - - - - - - - - -

(a) $7x - 28 = 0$; (b) $-\left(\dfrac{-28}{7}\right) = 4$

9. Suppose you rent a party hall for $50 plus $2 per person. After the party, you are billed for $130.

(a) Write a linear equation to represent the facts: _____

(b) How many attended? _____

- - - - - - - - - - - - - -

(a) $2x + 50 = 130$; (b) $x = 40$

LINEAR EQUATIONS IN TWO VARIABLES

We have been solving linear equations with one variable; they can always be solved by writing them in the form $ax + b = 0$ because $x = \dfrac{-b}{a}$. Now we are going to consider linear equations in two variables. These are not nearly as simple to solve.

10. A linear equation in one variable has the form $ax + b = 0$. A linear equation in two variables has the form $ax + by + c = 0$, where x and y are variables, a is the coefficient of x, b is the coefficient of y, and c is a constant value. $3x + 4y - 7 = 0$ is a linear equation in two variables. Refer to this example and answer the following questions.

(a) What is b? _____

(b) What is c? _____

(c) What is a? _____

_ _ _ _ _ _ _ _ _ _ _ _ _ _

(a) 4; (b) −7; (c) 3

11. A linear equation in two variables (or unknowns) has many possible solutions. To find a specific solution you need two equations written in the same two variables. The two are then called a <u>system</u> of linear equations and can be solved simultaneously.

Which of the following might be a system of linear equations in two variables?

___(a) $3x - 7 = 0$ and $5x - 8 = 0$
___(b) $3x + 2y = 9$ and $4x - y = 3$
___(c) $5x + 3 = 8y$ and $4 - 3y = 2x$

_ _ _ _ _ _ _ _ _ _ _ _ _ _

(b); (c) (Choice (a) has only one variable.)

SOLUTION BY SUBSTITUTION

12. A system of two linear equations may be solved by a method called <u>substitution</u>. In this method, you solve one equation for the value of one variable, then substitute that value in the other equation. For example, consider the following equations:

(1) $x - 2y = -8$

(2) $x + y = 7$

First solve equation (1) for the value of x:

$x = 2y - 8$

Then substitute this value of x in equation (2):

$(2y - 8) + y = 7$

Now simplify equation (2) and solve for y:

$3y = 15$

$y = 5$

Finally, take the value of y and substitute it into equation (1):

$x - 2(5) = -8$

$x = 2$

The solution to this system is $x = 2$, $y = 5$.

Follow the steps indicated below to solve the following system of linear equations:

① $x - 2y = -8$

② $2x - y = -1$

(a) Solve equation ① for x: x = _____

(b) Substitute (a) into equation ②: _____

(c) Simplify (b):   _____

(d) Solve equation ② for y: y = _____

(e) Substitute (d) into equation ①: _____

(f) Solve for x: x = _____

- - - - - - - - - - - - - - - - -

(a) $x = 2y - 8$; (b) $2(2y - 8) - y = -1$; (c) $3y = 15$; (d) 5;
(e) $x - 2(5) = -8$; (f) 2 (You can check your results any time by substituting the values of x and y into the original equations.)

13. The substitution method of solving two linear equations includes the following five steps:

Step 1: Solve one equation for one variable (either one will do).

Step 2: Substitute the expression for that variable into the other equation.

Step 3: Solve for the second variable.

Step 4: Substitute that solution into the first equation.

Step 5: Solve for the value of the first variable.

Use these five steps to solve the system below:

① $x + y = 7$

② $2x - y = -10$

(a) Solve equation ① for x: x = _____

(b) Substitute (a) into equation ②: _____

(c) Simplify (b):   _____

(d) Find y: y = _____

(e) Substitute y in equation ①: _____

(f) Find x: x = _____

- - - - - - - - - - - - - - - -

(a) $x = 7 - y$; (b) $2(7 - y) - y = -10$; (c) $-3y = -24$ (or $3y = 24$);
(d) $y = 8$; (e) $x + 8 = 7$; (f) $x = -1$

14. Solve the following system of two linear equations:

    (1) $x + y = 15$

    (2) $x - y = -1$

    (a) Solve equation (1) for x: $x = $ _____

    (b) Substitute (a) into equation (2): _____

    (c) Simplify (b):   _____

    (d) Find y: $y = $ _____

    (e) Substitute y in equation (1): _____

    (f) Solve for x: $x = $ _____

- - - - - - - - - - - - - - - -

(a) $x = 15 - y$; (b) $(15 - y) - y = -1$; (c) $-2y = -16$; (d) $y = 8$;
(e) $x + 8 = 15$; (f) $x = 7$

15. Solve the system of two linear equations below by solving for one
    variable, substituting it in the other equation, then substituting the
    result back into the first equation. Solve for y first.

    (1) $2x + y = 7$

    (2) $3x - y = 3$

-- -- -- -- -- -- -- -- -- --

x = 2, y = 3   Solution:

Solve ① for y: y = 7 − 2x

Substitute in ② : 3x − (7 − 2x) = 3

Simplify: 5x = 10

Solve for x: x = 2

Substitute x in ①: 2(2) + y = 7

Solve for y: y = 3

16. Solve the following system of linear equations by substitution.

$$x + 2y = -1$$
$$2x + 5y = -7$$

-- -- -- -- -- -- -- -- -- --

x = 9, y = -5   Solution: x = -1 − 2y

2(-1 − 2y) + 5y = -7

y = -5

x + 2(-5) = -1

x = 9

17. Solve the system of linear equations below by substitution.

$$x - 2y = -8$$
$$2x - \ y = -10$$

- - - - - - - - - - - - - -

$x = -4$, $y = 2$

SOLUTION BY COMPARISON

We will now look at another method of solving linear equations, this time by comparison.

18. Two linear equations in two variables can be solved simultaneously by the method of comparison, as shown below. When you compare the two equations, you notice that by adding or subtracting them, you can effectively eliminate one of the two variables.

    ① $x + y = 7$

    ② $2x - y = -1$

    Add the two linear equations ① and ②: $3x = 6$

    Solve for the obvious variable: $x = 2$

    Substitute this variable into equation ①: $2 + y = 7$

    Solve for the other variable: $y = 5$

This method can be checked by inserting both values into the second equation.

Now solve the system below using the method of comparison.

    ① $x - 2y = -8$

    ② $x + \;\; y = 7$

(a) Subtract ② from ① (remember that when you subtract you must change all the signs in ②): _____

(b) Solve for y: $y =$ _____

(c) Substitute the value of y in ①: _____

(d) Solve for x: $x =$ _____

- - - - - - - - - - - - - -

(a) $-3y = -15$;   (b) $y = 5$;   (c) $x - 2(5) = -8$;   (d) $x = 2$

19. Combining the equations so as to eliminate one variable is the first step in using the method of comparison to solve a system of two linear equations. This can often be done by simply adding or subtracting the two equations. However, it sometimes requires more. One of the equations may have to be multiplied by a constant amount before the addition or subtraction will eliminate one of the variables. Consider, for example, the following system of linear equations:

$$\text{(1)} \quad 2x - 5y = 6$$
$$\text{(2)} \quad 4x + y = 1$$

To solve this system, first multiply equation ① by 2, giving equation ③. Then proceed to subtract equation ③ from equation ②. Solve the system above using the steps listed below.

(a) Multiply ① by 2 to obtain ③: _____

(b) Subtract ③ from ②: _____

(c) Solve for y: y = _____

(d) Substitute the value of y in ①: _____

(e) Solve for x: x = _____

- - - - - - - - - - - - - - -

(a) $4x - 10y = 12$; (b) $11y = -11$; (c) $y = -1$; (d) $2x - 5(-1) = 6$; (e) $x = \frac{1}{2}$

20. In solving the system of linear equations below by the method of comparison, the <u>first</u> step is to:

$$\text{(1)} \quad 2x - 5y = 16$$
$$\text{(2)} \quad 3x + y = 7$$

___(a) add ① and ②

___(b) multiply ② by 5

___(c) multiply ① by 3

___(d) subtract ② from ①

- - - - - - - - - - - - - -

(b)

21. Solve this system of linear equations by the method of comparison.

    ① $2x - 5y = 16$

    ② $3x + y = 7$

(a) Multiply ② by 5 to obtain ③: _____

(b) Add ① and ③: _____

(c) Solve for x: x = _____

(d) Substitute x in ①: _____

(e) Solve for y: y = _____

- - - - - - - - - - - - - - -

(a) $15x + 5y = 35$; (b) $17x = 51$; (c) $x = 3$; (d) $2(3) - 5y = 16$;
(e) $y = -2$

22. Several systems of equations are given below. In each system, one equation must be multiplied by a constant before the equations are added or subtracted to eliminate one variable. Write the equation number, and the constant it should be multiplied by.

Systems	Equation	Constant
(a) ① $x + 2y = -1$ ② $2x + 5y = -7$	_____	_____
(b) ① $x + y = 3$ ② $3x + 4y = 10$	_____	_____
(c) ① $5x + 4y = -10$ ② $3x - 2y = 16$	_____	_____

- - - - - - - - - - - - - -

(a) ①, 2; (b) ①, 3 or 4; (c) ②, 2

23. The systems given below were examined in frame 22. Find x and y for each.

(a) ① $x + y = 3$ (b) ① $5x + 4y = -10$

    ② $3x + 4y = 13$ ② $3x - 2y = 16$

- - - - - - - - - - - - - -

(a) $x = -1$, $y = 4$;   (b) $x = 2$, $y = -5$

24. Sometimes it is necessary to multiply both equations by a constant when using the comparison method. Solve the system below.

(1) $2x + 3y = -9$ (Hint: Multiply (1) by 3, (2) by 2.)

(2) $3x + 4y = -10$

- - - - - - - - - - - - - -

$x = 6$, $y = -7$ Solution:

(3) $6x + 9y = -27$

(4) $6x + 8y = -20$

$$y = -7$$

$$2x + 3(-7) = -9$$

$$2x = 12$$

$$x = 6$$

LINEAR EQUATIONS IN MORE VARIABLES

By now you should be able to solve any system of two linear equations in two variables if the system has a solution. We have not discussed how to identify whether a system has a solution; that is beyond the scope of this book. We also have not shown that the solutions we have found are unique; that too is beyond our scope.

We will now go on to solve systems of more than two linear equations with more than two variables. This is usually done by using the method of comparison successively to eliminate variables.

25. The first step in using comparison to solve systems of equations is to eliminate one variable. Study the beginning steps below of this system of equations:

$$\text{(1)} \quad 2x + y - z = 10$$
$$\text{(2)} \quad x + 3y - z = 30$$
$$\text{(3)} \quad 3x + 3y - 2z = 25$$

Multiply ① by 2 to obtain ④: $4x + 2y - 2z = 20$

Subtract ④ from ③ to obtain ⑤: $-x + y \quad\quad = 5$

Multiply ② by 2 to obtain ⑥: $2x + 6y - 2z = 60$

Subtract ⑥ from ③ to obtain ⑦: $x - 3y \quad\quad = -35$

The next step clearly involves equations ⑤ and ⑦, giving the value of one variable as _____.

- - - - - - - - - - - - - - -

$-2y = -30$, or $y = 15$

26. Now that you know $y = 15$, you can get the value of x from equation ⑦: $x =$ _____.

- - - - - - - - - - - - - - -

10

27. Use your values of x and y to find the value of z in the system of frame 25.

- - - - - - - - - - - - - - -

$z = 25$ (Substitute x and y into ①, ②, or ③.)

A system of linear equations in three unknowns requires a minimum of three equations. A system in four unknown variables requires at least four equations, and so on. A new method of solving systems of equations that is directly applicable to many phases of computer programming will be introduced in the next chapter.

Now take the Self-Test on the next page.

SELF-TEST

This Self-Test will help you evaluate whether or not you have mastered the chapter objectives and are ready to go on to the next chapter.   Answer each question to the best of your ability.   Correct answers are given at the end of the test.

1. Which of the equations below are linear?

___(a) $3x^3 + 4x - 1 = 0$
___(b) $3w + 4x + y = 0$
___(c) $5x + 6 = 0$

2. Solve the linear equations given below.

(a) $5x = 4$

(b) $x - 10 = 4$

(c) $18x - 5 = 4$

3. Solve the system of linear equations below by substitution.

① $2x - 5y = 6$
② $4x + y = 1$

4. Solve the system of linear equations below by comparison.

① $5x + 6y = 2$
② $2x + 3y = -1$

ANSWERS TO SELF-TEST

Compare your answers to the Self-Test with the correct answers given below. If all of your answers are correct, you are ready to go on to the next chapter. If you missed any questions, study the frames indicated in parentheses following the answer.

1. (b); (c) (frames 1-2)

2. (a) $x = 4/5$; (b) $x = 14$; (c) $x = 1/2$ (frames 3-10)

3. $x = 1/2, y = -1$ Solution:

① $2x - 5y = 6$

② $4x + y = 1$

Solve ② for y: $y = 1 - 4x$

Substitute in ①: $2x - 5(1 - 4x) = 6$

$$22x = 11$$

$$x = 1/2$$

Substitute x in ②: $4(1/2) + y = 1$

$$2 + y = 1$$

$$y = -1 \quad \text{(frames 11-17)}$$

4. $x = 4, y = -3$ Solution:

① $5x + 6y = 2$

② $2x + 3y = -1$

Multiply ② by 2 to obtain ③: $4x + 6y = -2$

Subtract ③ from ①: $1x = 4$

$$x = 4$$

Substitute x in ①: $5(4) + 6y = 2$

$$6y = -18$$

$$y = -3$$

(frames 18-24)

CHAPTER ELEVEN
Matrix Algebra

Matrices are widely used in mathematics and computer programming, sometimes being referred to as tables or arrays. A matrix, table, or array is a rectangular arrangement of elements having two dimensions —length and width. Matrices are used in mathematics for solving systems of equations, in research for storing and processing quantities of raw data, and in business for such problems as cost accounting.

When you complete this chapter, you will be able to:

- solve a system of linear equations using determinants;

- use double subscripts in working with matrices;

- find the value of the determinant of a matrix;

- identify the function of the identity matrix for third-order matrices;

- identify the function of the inverse of a matrix.

In this chapter you will learn the bare essentials of matrix algebra. You may find additional information in Matrix Algebra by Richard C. Dorf (New York: John Wiley & Sons, 1969).

DETERMINANTS

A form of matrix called a <u>determinant</u> can be used to solve systems of linear equations. In this section, you will learn to solve systems of two linear equations using determinants. Then we will extend this knowledge to solving larger systems, both by hand and by computer.

In the last chapter, you solved systems of linear equations by two methods—comparison and substitution. In this chapter, you will learn a new way, but the notation is a bit different.

1. Below are two linear equations using a different kind of notation:

$$a_1x + b_1y = c_1$$

$$a_2x + b_2y = c_2$$

In the general equations above:

___(a) c_1 is a variable
___(b) x and y are variables with unknown values
___(c) $b_1y = b_2y$
___(d) a_1, b_1, and c_1 are coefficients with constant values

- - - - - - - - - - - - - - -

(b); (d)

2. Below are two general equations (on the left) and a system of linear equations with numeric coefficients (on the right):

$$a_1x + b_1y = c_1 \qquad\qquad x - 2y = -8$$

$$a_2x + b_2y = c_2 \qquad\qquad 2x - y = -1$$

For the system on the right, identify:

(a) b_2 _____

(b) c_1 _____

(c) a_2 _____

- - - - - - - - - - - - - -

(a) −1 (−1y = −y); (b) −8; (c) 2

3. Identify the items below for this system of linear equations.

$$2x + 3y = -9$$
$$3x + 4y = -10$$

(a) a_1 _____

(b) b_2 _____

(c) c_1 _____

- - - - - - - - - - - - - -

(a) 2; (b) 4; (c) -9

4. Use the constant values listed below on the right to construct a system of linear equations in the form of those on the left.

$$a_1x + b_1y = c_1$$
$$a_2x + b_2y = c_2$$

$a_1 = 2,\ b_1 = 1,\ c_1 = 7$

$a_2 = 3,\ b_2 = -1,\ c_2 = 3$

- - - - - - - - - - - - - -

$2x + y = 7;\ 3x - y = 3$

5. Using only the general formulas for a linear system in two variables, the method of comparison gives a formula for finding values of x and y directly. Study the procedure on page 244, then answer these questions.

(a) Write the formula for x in the system of general formulas:

_____

(b) Write the formula for y: _____

(c) Which is identical in both, the numerator (top part) or the denominator (bottom part)?

inator (bottom part)? _____

- - - - - - - - - - - - - -

(a) $x = \dfrac{c_1b_2 - c_2b_1}{a_1b_2 - a_2b_1}$; (b) $y = \dfrac{a_1c_2 - a_2c_1}{a_1b_2 - a_2b_1}$; (c) denominator

Procedure for Finding Values of x and y Directly

$$\text{(1)} \quad a_1x + b_1y = c_1$$
$$\text{(2)} \quad a_2x + b_2y = c_2$$

Multiply (1) by b_2:   (3) $a_1b_2x + b_1b_2y = c_1b_2$

Multiply (2) by b_1:   (4) $a_2b_1x + b_2b_1y = c_2b_1$

Subtract (4) from (3):   (5) $(a_1b_2 - a_2b_1)x = c_1b_2 - c_2b_1$

Multiply (2) by a_1:   (6) $a_1a_2x + a_1b_2y = a_1c_2$

Multiply (1) by a_2:   (7) $a_1a_2x + a_2b_1y = a_2c_1$

Subtract (7) from (6):   (8) $(a_1b_2 - a_2b_1)y = a_1c_2 - a_2c_1$

Divide (5) by $a_1b_2 - a_2b_2$
$(a_1b_2 - a_2b_1 \neq 0)$:   (9) $x = \dfrac{c_1b_2 - c_2b_1}{a_1b_2 - a_2b_1}$

Divide (8) by $a_1b_2 - a_2b_1$
$(a_1b_2 - a_2b_1 \neq 0)$:   (10) $y = \dfrac{a_1c_2 - a_2c_1}{a_1b_2 - a_2b_1}$

6. The formulas for finding x and y are neither easy to write nor easy to work with. For this reason, they are simplified. The square arrangement of elements below is called the <u>determinant definition</u>.

$$\begin{vmatrix} a_1 & b_1 \\ a_2 & b_2 \end{vmatrix} = a_1b_2 - a_2b_1$$

This determinant definition defines the

___(a) value of x for the general formulas
___(b) value of y for the general formulas
___(c) value of the common denominator of x and y

- - - - - - - - - - - - - - - -

(c)

VALUE OF A DETERMINANT

7. A column is vertical; a_1 and a_2 make up a column. A row is horizontal; a_2 and b_2 make up a row. Compare the following determinant definition on the left with the value on the right.

$$\begin{vmatrix} a_1 & b_1 \\ a_2 & b_2 \end{vmatrix} = a_1 b_2 - b_1 a_2$$

Which of the following statements describes the way the value on the right is reached?

___(a) Multiply together the elements in a column.
___(b) Multiply together the elements in a row.
___(c) Multiply together the elements in a diagonal.

- - - - - - - - - - - - - -

(c)

8. The determinant below has a value of -5.

$$\begin{vmatrix} a_1 & b_1 \\ a_2 & b_2 \end{vmatrix} = a_1 b_2 - b_1 a_2 \qquad \begin{vmatrix} 2 & 1 \\ 3 & -1 \end{vmatrix} = (2)(-1) - (1)(3) = -5$$

Find the values of the determinants below.

(a) $\begin{vmatrix} 1 & 2 \\ 2 & 1 \end{vmatrix}$ = _____

(b) $\begin{vmatrix} 2 & 5 \\ 4 & 1 \end{vmatrix}$ = _____

- - - - - - - - - - - - - -

(a) -3 (1 − 4); (b) -18 (2 − 20)

9. We have been discussing the determinant that is used as the common denominator of the variables in a system of linear equations. This is called the <u>determinant of the system</u>. Write the determinant of the system for the system below.

$$2x + 3y = -9$$

$$3x + 4y = -10$$

$$\begin{vmatrix} & \\ & \end{vmatrix} = \underline{\quad\quad} - \underline{\quad\quad} = \underline{\quad\quad}$$

- - - - - - - - - - - - - - - -

$$\begin{vmatrix} 2 & 3 \\ 3 & 4 \end{vmatrix} = 8 - 9 = -1$$

10. Notice that the values <u>a</u> and b in the determinant of a system are arranged just as they are in the system itself.

$$a_1 x + b_1 y = c_1$$

$$a_2 x + b_2 y = c_2$$

$$\begin{vmatrix} a_1 & b_1 \\ a_2 & b_2 \end{vmatrix}$$

Write the determinant of the system for this general system.

$$\begin{vmatrix} a_1 & b_1 \\ a_2 & b_2 \end{vmatrix} = \underline{\hspace{6cm}}$$

- - - - - - - - - - - - - -

$a_1 b_2 - a_2 b_1$ (or $a_1 b_2 - b_1 a_2$, or $b_1 a_2 - a_2 b_1$. The form shown here is most often used.)

11. The values of all determinants of this size (two rows by two columns, called two-by-two or second order) are figured in the same way. What would be the value of the following determinants?

(a) $\begin{vmatrix} c_1 & b_1 \\ c_2 & b_2 \end{vmatrix} = \underline{\hspace{3cm}}$
(b) $\begin{vmatrix} a_1 & c_1 \\ a_2 & c_2 \end{vmatrix} = \underline{\hspace{3cm}}$

- - - - - - - - - - - -

(a) $c_1b_2 - c_2b_1$; (b) $a_1c_2 - a_2c_1$

12. On page 244 we showed that $x = \dfrac{c_1b_2 - c_2b_1}{a_1b_2 - a_2b_1}$. This is shown in determinant form below:

$$x = \frac{\begin{vmatrix} c_1 & b_1 \\ c_2 & b_2 \end{vmatrix}}{\begin{vmatrix} a_1 & b_1 \\ a_2 & b_2 \end{vmatrix}}$$

If $y = \dfrac{a_1c_2 - a_2c_1}{a_1b_2 - a_2b_1}$, write y in determinant form below.

- - - - - - - - - - - -

$$y = \frac{\begin{vmatrix} a_1 & c_1 \\ a_2 & c_2 \end{vmatrix}}{\begin{vmatrix} a_1 & b_1 \\ a_2 & b_2 \end{vmatrix}}$$

13. Use the determinant quotient on the left to find the value of x in the system on the right, using the steps listed below.

$$x = \frac{\begin{vmatrix} c_1 & b_1 \\ c_2 & b_2 \end{vmatrix}}{\begin{vmatrix} a_1 & b_1 \\ a_2 & b_2 \end{vmatrix}}$$

$2x + 3y = -9$

$3x + 4y = -10$

(a) Construct the numerator and denominator determinants, substituting values from the given system:

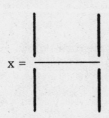

$$x = \frac{\begin{vmatrix} & \end{vmatrix}}{\begin{vmatrix} & \end{vmatrix}}$$

(b) Find the numerator value: _____

(c) Find the denominator value: _____

(d) Find $x = \dfrac{\text{numerator}}{\text{denominator}}$: _____

- - - - - - - - - - - - - -

(a) $\dfrac{\begin{vmatrix} -9 & 3 \\ -10 & 4 \end{vmatrix}}{\begin{vmatrix} 2 & 3 \\ 3 & 4 \end{vmatrix}}$

(b) -6 $(-36 + 30)$; (c) -1 $(8 - 9)$; (d) $x = 6$ $(-6/-1 = 6)$

14. The determinant for the value of y has the same denominator as that for x. Therefore, this part need not be recalculated. What is the value of y for the system given in frame 13?

$$y = \frac{\begin{vmatrix} a_1 & c_1 \\ a_2 & c_2 \end{vmatrix}}{\begin{vmatrix} a_1 & b_1 \\ a_2 & b_2 \end{vmatrix}}$$

- - - - - - - - - - - - - - -

y = -7   Solution:

$$\begin{vmatrix} 2 & -9 \\ 3 & -10 \end{vmatrix}$$ numerator = -20 + 27 = 7

$$\begin{vmatrix} 2 & 3 \\ 3 & 4 \end{vmatrix}$$ denominator (from frame 13) = -1

$$y = \frac{7}{-1} = -7$$

15. We noted earlier that the determinant of the system (common denominator determinant) had exactly the same arrangement of $\underline{a}$'s and b's as did the system. Examine the numerators below:

$$x = \frac{\begin{vmatrix} c_1 & b_1 \\ c_2 & b_2 \end{vmatrix}}{\begin{vmatrix} a_1 & b_1 \\ a_2 & b_2 \end{vmatrix}} \qquad y = \frac{\begin{vmatrix} a_1 & c_1 \\ a_2 & c_2 \end{vmatrix}}{\begin{vmatrix} a_1 & b_1 \\ a_2 & b_2 \end{vmatrix}}$$

(a) In the numerator of the formula for x, the values of a_1 and a_2 that appear in the determinant of the system are replaced by

_____.

(b) In the numerator of the formula for y, the values of b_1 and b_2 that appear in the determinant of the system are replaced by

_____.

- - - - - - - - - - - - -

(a) c_1 and c_2; (b) c_1 and c_2

16. Match the determinants below with their positions in formulas for x and y.

___(a) $\begin{vmatrix} a_1 & b_1 \\ a_2 & b_2 \end{vmatrix}$

___(b) $\begin{vmatrix} c_1 & b_1 \\ c_2 & b_2 \end{vmatrix}$

___(c) $\begin{vmatrix} a_1 & c_1 \\ a_2 & c_2 \end{vmatrix}$

1. numerator for x

2. numerator for y

3. denominator for x

4. denominator for y

- - - - - - - - - - - -

(a) 3, 4; (b) 1; (c) 2

17. Solve this system using the steps below.

$$x - 2y = -8$$
$$x + y = 7$$

(a) Write the determinant fractions for x and y, using values from the system given above:

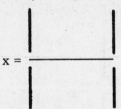   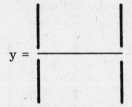

(b) Solve for x: _____

(c) Solve for y: _____

- - - - - - - - - - - - - -

(a) $x = \dfrac{\begin{vmatrix} -8 & -2 \\ 7 & 1 \end{vmatrix}}{\begin{vmatrix} 1 & -2 \\ 1 & 1 \end{vmatrix}}$ $y = \dfrac{\begin{vmatrix} 1 & -8 \\ 1 & 7 \end{vmatrix}}{\begin{vmatrix} 1 & -2 \\ 1 & 1 \end{vmatrix}}$

(b) $x = 2$ $\left(\dfrac{-8 + 14}{1 + 2}\right)$; (c) $y = 5$ $\left(\dfrac{7 + 8}{3}\right)$

THREE-BY-THREE DETERMINANTS

You can now use two-by-two (or second order) determinants for solving two linear equations in two variables simultaneously. Now we will consider larger systems in more variables, and consequently higher-order determinants.

18. A system of three linear equations in three variables can be written in the same notation as before:

$$a_1x + b_1y + c_1z = d_1$$
$$a_2x + b_2y + c_2z = d_2$$
$$a_3x + b_3y + c_3z = d_3$$

The determinant of the system, as before, has the same arrangement of a's, b's, and c's as does the system itself. Write the determinant of the system.

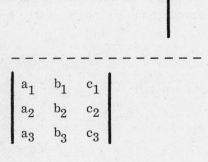

$$\begin{vmatrix} a_1 & b_1 & c_1 \\ a_2 & b_2 & c_2 \\ a_3 & b_3 & c_3 \end{vmatrix}$$

19. Below is the definition for third-order determinants:

$$\begin{vmatrix} a_1 & b_1 & c_1 \\ a_2 & b_2 & c_2 \\ a_3 & b_3 & c_3 \end{vmatrix} = a_1 b_2 c_3 + a_2 b_3 c_1 + a_3 b_1 c_2 - a_3 b_2 c_1 - a_2 b_1 c_3 - a_1 b_3 c_2$$

This definition can be clarified by the following diagram.

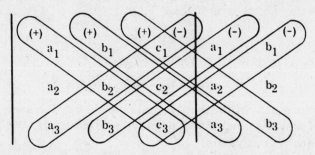

As the diagram shows, you can add the products of the three elements on each of the diagonals sloping down to the right. Subtract from this sum the products of the diagonals sloping down to the left.

Use the information above to complete the definition of the determinant below.

$$\begin{vmatrix} d_1 & b_1 & c_1 \\ d_2 & b_2 & c_2 \\ d_3 & b_3 & c_3 \end{vmatrix} = d_1 b_2 c_3 + b_1 c_2 d_3 + \underline{\hspace{1.5cm}} - b_1 d_2 c_3 - d_1 c_2 b_3 - \underline{\hspace{1cm}}$$

- - - - - - - - - - - - - - - -

$c_1 d_2 b_3; \; c_1 b_2 d_3$

20. The determinant in the last frame has the constant d in place of the constant $\underline{a}$. Consider the arrangement of values in the second-order determinants. The position of this determinant is probably:

___(a) a denominator
___(b) numerator of x
___(c) numerator of y
___(d) numerator of z

- - - - - - - - - - - - - - -

(b)

MATRICES

We will not, at this time, go into the solution of third order and higher determinants by hand. Instead, we shall now introduce a different notation that uses two subscripts for each variable. This is more useful in computer work with matrices, of which determinants are just a part.

21. The constant values in a system of linear equations may be considered a matrix.*

$$\begin{pmatrix} a_1 & b_1 & c_1 \\ a_2 & b_2 & c_2 \end{pmatrix} \quad \text{or} \quad \begin{pmatrix} a_{11} & a_{12} & a_{13} \\ a_{21} & a_{22} & a_{23} \end{pmatrix}$$

The matrix on the left above differs from a determinant in that:

___(a) determinants are square
___(b) matrices are square
___(c) it has two subscripts for each variable

– – – – – – – – – – – – – –

(a)

22. The circled element below is in the:

$$a_{11} \quad \boxed{a_{12}} \quad a_{13}$$
$$a_{21} \quad a_{22} \quad a_{23}$$

___(a) first row
___(b) first column
___(c) second row
___(d) second column

– – – – – – – – – – – – – –

(a); (d)

23. The notation a_{12} has a double subscript. The first subscript, 1, refers to the row. The second subscript, 2, refers to the column.

(a) a_{23} is in the _____ row

(b) a_{23} is in the _____ column

– – – – – – – – – – – – – –

(a) second; (b) third

_____

*We have used large parentheses to differentiate matrices from determinants.

24. The first subscript in any double-subscripted variable refers to the row; the second refers to the column (a_{rc}). You might have a variable value called m_{74}. This tells you the value is in the

_____ row and the _____ column.

- - - - - - - - - - - - - - - -

seventh; fourth

25. In one sense, a matrix is a table, as shown below:

column row	1	2	3	4
1	1	2	-1	-3
2	2	-1	1	5
3	3	2	-2	-3

Using the table as data, give the values of the double-subscripted variables below.

(a) a_{21} _____

(b) a_{32} _____

(c) a_{13} _____

(d) a_{31} _____

- - - - - - - - - - - - - - -

(a) 2; (b) 2; (c) -1; (d) 3

26. In a variable with two subscripts (called a double-subscripted variable), the first subscript always refers to the _____.

- - - - - - - - - - - - - -

row

27. A matrix of data is stored and processed by a computer either row-by-row or column-by-column, in order of subscripts. In row-by-row processing, a_{12} comes before a_{13}, which precedes a_{21}. Which representation below shows that all of row 1 is processed, then all of row 2?

___(a) a_{11} a_{12} a_{13} a_{21} a_{22} a_{23}

___(b) a_{11} a_{21} a_{12} a_{22} a_{13} a_{23}

___(c) a_{11} a_{22} a_{12} a_{21} a_{13} a_{23}

- - - - - - - - - - - - -

(a) (all of row 1, then all of row 2, etc.)

28. The subscripts of a variable are integers. The subscripts vary, however, with each element in the matrix. The usual symbols for subscripts are I, J, and K. In most matrices (two-dimensional arrays) only two subscripts will be needed. I and J are the variables most often used to represent the subscripts. The subscripted variable A would be written A_{IJ}.

In the notation A_{IJ}, which letter refers to the row of the element? _____

- - - - - - - - - - - - - -

I

29. Suppose you want to add up all the elements in a three-by-four matrix. Examine the flowchart segment below. This flowchart would be used to add up the elements of:

___(a) the entire matrix
___(b) any one column
___(c) a row

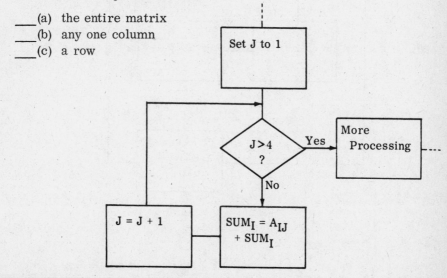

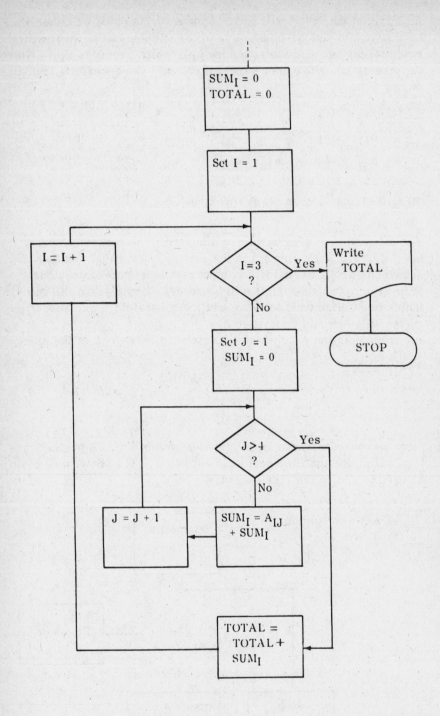

(c) (The row variable, I, remains the same. One element from each column is added in this row.)

30. Examine the flowchart segment on page 256. You are dealing with a three-by-four matrix:

$$
\begin{array}{cccc}
1 & 2 & -1 & -3 \\
2 & -1 & 1 & 5 \\
3 & 2 & -2 & -3
\end{array}
$$

(a) What row is added first? _____

(b) What is the value of TOTAL when I is first equal to 2? _____

(c) How many times will the block "I = I + 1" be performed? _____

(a) first; (b) -1; (c) 2 (I = 1 + 1 and I = 2 + 1)

We will not delve more deeply into the intricacies of double-subscripting. The important items to remember are:

- The first subscript indicates a row.
- The second subscript indicates a column.
- Double subscripts usually indicate a matrix.
- Processing is done in row or column order.

OPERATIONS ON MATRICES

31. Two matrices can be added together if they are the same size and shape. Corresponding elements are added, $a_{11} + b_{11}$, $a_{12} + b_{12}$, etc., in row order. Two matrices are given below.

$$
\begin{pmatrix} 6 & -1 & 4 \\ 9 & 2 & 3 \end{pmatrix} + \begin{pmatrix} 4 & 8 & -1 \\ 0 & -4 & 2 \end{pmatrix}
$$

Give the resulting matrix when these are added.

- - - - - - - - - - - -

$$\begin{pmatrix} 10 & 7 & 3 \\ 9 & -2 & 5 \end{pmatrix}$$

32. A matrix can be multiplied by a constant simply by multiplying each element in it by that constant. Multiply the matrix below by 3.

$$\begin{pmatrix} 4 & 8 & -1 \\ 0 & -4 & 2 \end{pmatrix}$$

- - - - - - - - - - - - -

$$\begin{pmatrix} 12 & 24 & -3 \\ 0 & -12 & 6 \end{pmatrix}$$

The techniques of multiplication of a matrix by another matrix is beyond the scope of this book. However, a few related facts are important to anyone in computer fields.

33. The matrix below is the <u>identity matrix</u> for any three-by-three matrix. When the identity matrix is multiplied by any three-by-three matrix, the product is the three-by-three matrix.

$$\begin{pmatrix} 1 & 0 & 0 \\ 0 & 1 & 0 \\ 0 & 0 & 1 \end{pmatrix}$$

Which of the multiplications below might be correct?

___(a) $$\begin{pmatrix} 4 & 1 & 4 \\ -1 & -2 & -1 \\ 2 & -1 & -2 \end{pmatrix} \begin{pmatrix} 1 & 0 & 0 \\ 0 & 1 & 0 \\ 0 & 0 & 1 \end{pmatrix} = \begin{pmatrix} 4 & 1 & 4 \\ -1 & -2 & -1 \\ 2 & -1 & -2 \end{pmatrix}$$

___(b) $$\begin{pmatrix} 1 & 0 & 0 \\ 0 & 1 & 0 \\ 0 & 0 & 1 \end{pmatrix} \begin{pmatrix} 1 & 0 & 0 \\ 0 & 1 & 0 \\ 0 & 0 & 1 \end{pmatrix} = \begin{pmatrix} 1 & 0 & 0 \\ 0 & 1 & 0 \\ 0 & 0 & 1 \end{pmatrix}$$

(continued on next page)

___(c) $\begin{pmatrix} 3 & 4 & 2 & 4 \\ 3 & 4 & 1 & 2 \\ 3 & 4 & 3 & -1 \\ 3 & 1 & 0 & 2 \end{pmatrix} \begin{pmatrix} 1 & 0 & 0 \\ 0 & 1 & 0 \\ 0 & 0 & 1 \end{pmatrix} = \begin{pmatrix} 3 & 4 & 2 & 4 \\ 3 & 4 & 1 & 2 \\ 3 & 4 & 1 & -1 \\ 3 & 1 & 0 & 2 \end{pmatrix}$

- - - - - - - - - - - - - - -

(a); (b) (Choice (c) shows a four-by-four matrix being multiplied by the three-by-three identity matrix.)

34. The <u>inverse</u> of a matrix is rather like the reciprocal of a number. A number times its reciprocal is equal to the identity element 1 (e.g., 6 x 1/6 = 1). A matrix times its inverse would be equal to

the _____ matrix.

- - - - - - - - - - - - - -

identity (Although matrix multiplication is beyond the scope of this book, an example is shown below of a matrix times its inverse, resulting in the identity matrix.)

$$\begin{pmatrix} 4 & 0 & 5 \\ 0 & 1 & -6 \\ 3 & 0 & 4 \end{pmatrix} \begin{pmatrix} 4 & 0 & -5 \\ -18 & 1 & 24 \\ -3 & 0 & 4 \end{pmatrix} = \begin{pmatrix} 1 & 0 & 0 \\ 0 & 1 & 0 \\ 0 & 0 & 1 \end{pmatrix}$$

35. Some square matrices do not have inverses. If the value of the determinant of a matrix is zero, then the corresponding matrix has no inverse. You can find the value of a two-by-two determinant. Which of the following matrices has an inverse?

___(a) $\begin{pmatrix} 1 & -1 \\ -1 & 1 \end{pmatrix}$ Determinant value   Is there an inverse?

        _____   _____

___(b) $\begin{pmatrix} 2 & -3 \\ 4 & -6 \end{pmatrix}$

        _____   _____

___(c) $\begin{pmatrix} 0 & 1 \\ 1 & 0 \end{pmatrix}$

        _____   _____

- - - - - - - - - - - - -

(a) 0, no; (b) 0, no; (c) -1, yes

The calculation of the inverse of a matrix involves finding out what matrix, when multiplied by the given matrix, produces the identity matrix. We will not go into this process here. The identity and inverse matrices were introduced very briefly. When necessary, you can refer to the book mentioned on page 241 for additional information.

In this chapter you learned to use matrices and their determinants in solving systems of linear equations. You have seen how double subscripts are used to refer to elements of a matrix or array. You have seen how rows and columns of matrices can be processed through flowcharting.

Now take the Self-Test on the next page.

SELF-TEST

This Self-Test will help you evaluate whether or not you have mastered the chapter objectives and are ready to go on to the next chapter. Answer each question to the best of your ability. Correct answers are given at the end of the test.

1. Use the values below to write a system of two linear equations in the form $ax + by = c$.

$$a_1 = 3 \qquad c_1 = 3 \qquad b_2 = 1$$
$$b_1 = -1 \qquad a_2 = 2 \qquad c_2 = 7$$

2. The value of the determinant below is _____.

$$\begin{vmatrix} 3 & -1 \\ 2 & 1 \end{vmatrix}$$

3. In its general form, $x = \dfrac{c_1 b_2 - c_2 b_1}{a_1 b_2 - a_2 b_1}$. Write this in determinant form.

4. The determinant expression for x is identical to that for y in the

    ___(a) numerator
    ___(b) denominator

5. Rewrite the determinant below using double subscripts for a matrix.

$$\begin{pmatrix} a_1 & b_1 & c_1 \\ a_2 & b_2 & c_2 \\ a_3 & b_3 & c_3 \end{pmatrix} = \begin{pmatrix} a_{11} & & \\ & & \\ & & \end{pmatrix}$$

6. In the notation A_{IJ}, which subscript refers to the column number?

    _____

7. The inverse of a square matrix is:

    ___(a) its identity matrix
    ___(b) similar to the reciprocal of a number
    ___(c) found by multiplying the matrix times the identity matrix

ANSWERS TO SELF-TEST

Compare your answers to the Self-Test with the correct answers given below. If all of your answers are correct, you are ready to go on to the next chapter. If you missed any questions, study the frames indicated in parentheses following the answer.

1. $3x - y = 3$
   $2x + y = 7$ (frames 1-4)

2. $5 \ [3 - (-2)]$ (frames 6-11)

3. $\dfrac{\begin{vmatrix} c_1 & b_1 \\ c_2 & b_2 \end{vmatrix}}{\begin{vmatrix} a_1 & b_1 \\ a_2 & b_2 \end{vmatrix}}$ (frames 11-13)

4. (b)   (frames 5, 8-10)

5. $\begin{pmatrix} a_{11} & a_{12} & a_{13} \\ a_{21} & a_{22} & a_{23} \\ a_{31} & a_{32} & a_{33} \end{pmatrix}$ (frames 23-25)

6. J (second)   (frames 22-23)

7. (b)   (frames 34-35)

CHAPTER TWELVE

Game Theory

The mathematics of game theory is closely related to that of linear systems and matrices. Probability theory, too, is involved. In game theory, situations are considered in which two (or more) people act, and their actions influence, but do not determine, the outcome of a single event. Each player tries to act so that he comes out ahead. Naturally, their strategies may conflict. Examples of such games are tic-tac-toe, checkers, chess, bridge, and poker.

In a larger sense, the principles of game theory can be applied to life and to decision-making done in many phases of business, science, research, and the military. The fundamental value of game theory lies in its application to basic methods of decision theory.

When you complete this chapter, you will be able to:

- identify a two-by-two matrix game as strictly determined or not strictly determined;

- identify a strictly determined game as fair or biased;

- identify each player's optimal strategy in a strictly determined game;

- identify a mixed strategy as optimal in a nonstrictly determined game.

A SIMPLE MATRIX GAME

For the first sequence of frames in this chapter, we will construct a very simple game using four cards. Consider two players, R and C. R has two cards, a red 7 and a black 7. C has two cards, a red 9 and a black 7. When a bell sounds, each player exposes just one of his cards. Refer back to this explanation as necessary.

1. If the cards played are the same color, R wins the dollar difference between the numbers on the cards. If the cards played are not the same color, C wins the dollar difference between the numbers on the cards. Suppose R plays a red 7 and C plays a red 9.

    (a) Who wins? _____

    (b) How much? _____

    – – – – – – – – – – – – – – –

    (a) R wins (The cards are the same color.); (b) $2

2. Suppose C plays a red 9 and A plays a black 7.

    (a) Who wins? _____

    (b) How much? _____

    – – – – – – – – – – – – – –

    (a) C; (b) $2

3. Suppose R and C both play a black 7.

    (a) Who wins? _____

    (b) How much? _____

    – – – – – – – – – – – – – –

    (a) R (the colors match); (b) $0 (no difference in number)

4. The results of the game in frame 3 can be represented in a table and considered to be a matrix.

		C	
		Black 7	Red 9
R	Black 7	0	-2
	Red 7	0	+2

This table shows the amounts won from the point of view of which player? _____

- - - - - - - - - - - - - -

R (Note that R represents rows and C represents columns.)

5. In which of the following cases does player C win money (refer to the table in frame 4)?

___(a) C plays black 7, R plays black 7
___(b) C blays black 7, R plays red 7
___(c) C plays red 9, R plays black 7
___(d) C plays red 9, R plays red 7

- - - - - - - - - - - - - -

(c) (-2 in the matrix means R loses $2; thus, C wins.)

6. Suppose R plays his black 7. What will R win if C plays his (refer to the table in frame 4):

(a) black 7? _____

(b) red 9? _____

- - - - - - - - - - - - - -

(a) 0 (nothing); (b) -2 (R will lose $2)

7. Suppose R plays his other card, his red 7. What will he win if C plays his:

(a) black 7? _____

(b) red 9? _____

- - - - - - - - - - - - - -

(a) 0; (b) $2

8. You have seen what player R would win in each of four events. If you were R, what card would you play for highest winnings?

    ___(a) alternate, starting with red 7
    ___(b) always play red 7
    ___(c) always play black 7
    ___(d) play red 7 three times, then black 7 once

- - - - - - - - - - - - - -

(b) (R cannot lose, and may sometimes win, if he always plays his red 7.)

9. Consider player C.

(a) If C wins $2, he must have played his _____.

(b) If C loses $2, he must have played his _____.

(c) To be certain he will not lose, C must play his _____.

- - - - - - - - - - - - - -

(a) red 9; (b) red 9; (c) black 7

10. Suppose R and C play this game ten times. Each time R plays his red 7. The first time C plays his red 9.

(a) The next nine times, C plays his black 7. Who wins? _____

(b) How much does he win? _____

- - - - - - - - - - - - - -

(a) R; (b) $2 (The last nine times resulted in $0.)

11. R's decision to "play the red 7" is called a <u>strategy</u>. This strategy assures R that the smallest amount he can win is:

    ___(a) – $2
    ___(b) $0
    ___(c) $2

- - - - - - - - - - - - - -

(b) (R cannot lose with this strategy.)

12. C's original strategy in the game of frame 10 was:

    ___(a) play red 9
    ___(b) play red 7
    ___(c) play black 7

- - - - - - - - - - - - - -

(a)

13. C lost $2 using his first strategy. Suppose he used the strategy again and had no way of knowing what R would play. What would be the probability that C would win? _____

- - - - - - - - - - - - - -

1/2 (R could play one of two cards.)

14. Refer back to frame 10. What was C's final strategy?

_____

- - - - - - - - - - - - - -

play black 7 (play to break even)

STRATEGIES FOR STRICTLY DETERMINED GAMES

15. We have seen that the matrix below represents a fair game; neither player will win from the other. It represents a <u>strictly determined game</u>.

0	-2
0	2

Any two-by-two matrix represents a strictly determined game if it contains one entry that is both equal to the minimum of its row and the maximum of its column. Which entry above contains this value, v? _____

- - - - - - - - - - - - - -

0   [The lower lefthand 0 is the minimum of row (0   2) and the maximum of column (0   0).]

16. In a strictly determined game, each player can figure out the other's best strategy. Thus he knows what the other will (or should) do. But first a player must know if the game is strictly determined. If it is, there will be a value in the matrix that is the minimum of its row and the maximum of its column. Which of the two-by-two matrices below is a strictly determined game?

___(a)

0	1
-3	10

___(b)

0	1
2	0

___(c)

5	2
-7	-4

- - - - - - - - - - - - - - -

(a); (c) [In (a), 0 is the minimum of its row (0 1) and the maximum of its column (0 -3). In (c), 2 is the minimum of its row (5 2) and the maximum of its column (2 -4). In (b), no element fulfills both conditions.]

17. A strictly determined game is said to be fair if neither player wins from the other. The matrix also gives you this information. If the determining value v is 0, the game is fair; neither player wins or loses. Examine the strictly determined games in frame 16. Which

    one is fair? _____

- - - - - - - - - - - - - - -

(a) (v = 0 in this matrix)

18. Which of the games below is strictly determined? _____

    Which is fair? _____

    (a)

0	2
-1	4

    (b)

3	1
-4	0

    (c)

5	0
0	2

- - - - - - - - - - - - - - -

strictly determined: (a) and (b)
fair: (a)

19. The best strategy for a player in a strictly determined game depends on whether he is in the R or C position. R should play the row that contains v. C should play the column that contains v. Consider this game:

4	2
-5	0

    (a) Is it fair? _____

    (b) What is R's best strategy? _____

    (c) What is C's best strategy? _____

- - - - - - - - - - - - - - -

(a) no (v ≠ 0); (b) play the first row; (c) play the second column

20. The entries in a game matrix are from the point of view of R. Consider again the game of frame 19. C's best strategy is to play the second column. If he does this, he might win as much as $0 and lose as much as $2. If C were to play the other column, he might:

(a) win as much as _____

(b) lose as much as _____

- - - - - - - - - - - - - - -

(a) $5; (b) $4

21. In frame 19, you decided that R's best strategy was to play the first row. If he does this, he cannot lose anything. All C can do is keep his losses to a minimum. What would happen if C decided to take a chance and play the first column? _____

_____

- - - - - - - - - - - - - - -

C would lose $4 since R will always play row 1.

You have seen that some games are strictly determined and that some of these strictly determined games are fair. You have seen that there is a best strategy for either player in these games, and you have learned to identify that strategy. Now we shall go on to those games that are not strictly determined.

STRATEGIES FOR NONSTRICTLY DETERMINED GAMES

22. Which of the games below are not strictly determined?

___(a)

5	0
0	2

___(b)

0	4
0	2

___(c)

1	-1
-1	1

- - - - - - - - - - - - - - -

(a); (c) (In (b), v = 0; both zeros are the minimum of the row and the maximum of the column.)

23. A nonstrictly determined game can be identified from the matrix by noting the entries on the diagonals. Both of the entries on one diagonal must be greater than both of the entries on the other diagonal.

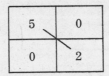

In the example above, 5 and 2 are both greater than 0 and 0. In choice (c) in the preceding frame, what are the entries on the greater valued diagonal? _____ and _____

- - - - - - - - - - - - - -

1 and 1

24. Let us consider the card game with which we began this unit. The rules of play are the same, but now R has a red 7 and a black 9, while C has a black 7 and a red 9. If the cards played are the same color, R wins; if the colors do not match, C wins the dollar difference. Fill in the values below. Remember to use R's point of view.

	C	
	Black 7	Red 9
R Black 9		
R Red 7		

- - - - - - - - - - - - -

2	0
0	2

25. The game of frame 24 is clearly not strictly determined since the 2's are greater than the 0's. The problem is finding each player's best strategy. The first step is to convince yourself that no one choice is clearly better for either player. For a single play the only key is secrecy. There is no difference as long as the opponent doesn't guess what you play. Letting the opponent guess what you are going to play would be more costly for _____.

- - - - - - - - - - - - - - -

C (R can't lose; C could break even at best.)

26. Referring again to the game described in frame 24, in order to op-
timize his results over a series of games, R should:

___(a) always play the black 9
___(b) always play the red 7
___(c) alternate the two cards
___(d) play one of two randomly

- - - - - - - - - - - - - - -

(d) (If C recognizes a pattern, R will win less.)

The best strategy for both R and C in this nonstrictly determined game is
a mixed strategy. In this case it is mixed $\frac{1}{2}$ to $\frac{1}{2}$ since the probability of
each simple event on the opponent's side is $\frac{1}{2}$.

  Formulas exist for calculating optimal strategies for nonstrictly determined games. These are based on probabilities and the values in the
matrix. With them, the optimal strategy and the amount of bias in favor
of one side can be calculated. For further information, refer to Finite
Mathematics with Business Applications by Kemeny, Schleifer, Snell,
and Thompson (New York: Prentice-Hall, 1962).

ZERO-SUM GAME

Now we will examine briefly one other type of game. This one involves
the matrix below.

		1	2	3	4	5	6
	1	6	2	5	4	3	1
	2	9	0	7	8	3	1
	3	7	3	5	4	2	3
R	4	2	5	1	8	6	0
	5	1	4	7	5	3	2
	6	8	4	9	0	6	7

(G across the top; R down the side.)

To play this game, you begin with a total score of zero. Then you roll two dice, one red and one green (R and G). You increase your score by the sum of the numbers in the red row, and you decrease it by the sum of the numbers in the green column. Suppose, for example, you roll a red 5 and a green 1. The numbers in the 5 row add up to 22. The numbers in the 1 column add up to 33. On this turn your score would be 22 − 33 = -11. You add the sum of the R row, then subtract the sum of the G column.

27. Suppose you roll a red 3 and a green 4. What is your score (refer to the rules above)? _____

- - - - - - - - - - - - - -

-5 (24 − 29)

28. If you roll a green 4 and a red 6, your score is _____ .

- - - - - - - - - - - - - -

5 (34 − 29)

29. The elements in the matrix can be subscripted A_{RG}. The flowchart below is used to find the:

___(a) value of G
___(b) sum of the elements in a row
___(c) sum of the elements in a column

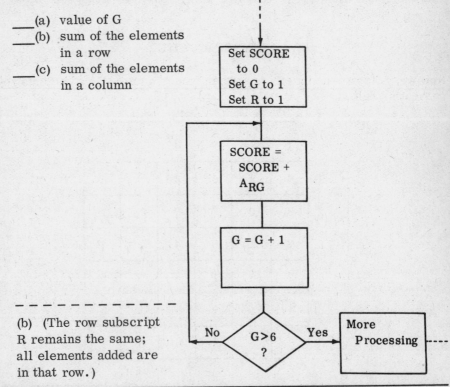

Set SCORE to 0
Set G to 1
Set R to 1

SCORE = SCORE + A_{RG}

G = G + 1

G > 6 ?   No   Yes

More Processing

(b) (The row subscript
R remains the same;
all elements added are
in that row.)

The flowchart on the right shows the programming required to compute scores with this six-by-six matrix game. In section 1 of the flowchart, the values of R and G (dice results) and the matrix are input. Then variables PLUS and MINUS (for sums of the row and column) are set to 0, and I and J are set to 1. In section 2, the sum of the R row is calculated as PLUS. In section 3, the sum of the G column is calculated as MINUS. Finally, in section 4, SCORE is calculated as PLUS − MINUS, and printed.

This flowchart could also be used for similar games with larger matrices of random numbers. You would simply change the 6's in the decision blocks to correspond with the size of your matrix.

You have now seen a very small part of game theory and how computers are used in relation to it. Game theory and its applications are growing increasingly important in the computer world.

Now take the Self-Test on the next page.

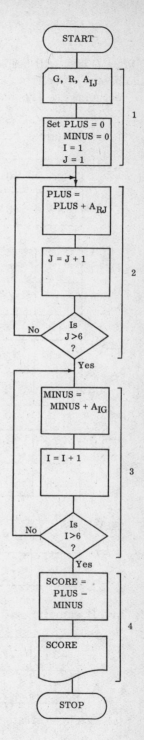

SELF-TEST

This Self-Test will help you evaluate whether or not you have mastered the chapter objectives. Answer each question to the best of your ability. Correct answers are given at the end of the test.

1. Match the games below with the descriptive phrases on the right.

___(a)

0	0
3	-3

1. strictly determined, not fair

2. strictly determined, fair

3. not strictly determined

___(b)

7	0
-2	6

___(c)

4	-3
1	2

___(d)

0	-1
7	1

___(e)

3	2
1	-1

2. In this game:

3	2
1	-1

(a) R's best strategy is to play row _____

(b) C's best strategy is to play column _____

3. In this game, R's best strategy is:

7	1
-6	5

___(a) always play row 1
___(b) always play row 2
___(c) mixed according to probabilities

ANSWERS TO SELF-TEST

Compare your answers to the Self-Test with the correct answers given below. If you missed any questions, study the frames indicated in parentheses following the answer.

1. (a) 2

   (b) 3

   (c) 3

   (d) 1

   (e) 1   (frames 15-23)

2. (a) 1

   (b) 2   (frames 15-20)

3. (a)   (frames 24-26)

Final Test

1. Find the decimal equivalents of:

   (a) 1100101_2 _____

   (b) 704.1_8 _____

   (c) BED_{16} _____

2. Find the binary equivalents of:

   (a) $200.$ _____

   (b) 704.1_8 _____

   (c) BED_{16} _____

3. Solve these problems.

   (a) 1100_2 (b) $B7\,39_{16}$ (c) $B7\,39_{16}$

       $x\ \ 110_2$ $+\,1FE0_{16}$ $-\,1FE0_{16}$

4. Assume that p, q, and r are true propositions. Give the truth value of each of the following.

   (a) $p \wedge \sim q$ _____

   (b) $p \vee \sim q$ _____

   (c) $p \vee (q \wedge \sim r)$ _____

   (d) $p \rightarrow \sim q$ _____

   (e) converse of $p \rightarrow \sim q$ _____

   (f) contrapositive of $p \rightarrow \sim q$ _____

5. Examine the flowchart on page 279. Suppose you have two input
    cards with the information shown below.

	Credit-code	Amount	Name
card 1	3	100.00	Harry Freeman
card 2	1	50.00	Angela Hoogterp

    Trace this information through the flowchart to find the output infor-
    mation for each of the input cards.

    (a) card 1 _____

    (b) card 2 _____

6. Solve these problems.

    (a) $(0.373 \times 10^6) + (0.19 \times 10^4)$

    (b) $\begin{array}{r} 0.67 \times 10^2 \\ \times\ 0.007 \times 10^6 \\ \hline \end{array}$

7. At 6 percent interest, compounded semiannually, how much interest
    would you earn on $700 in two years?

8. Write a series of four terms whose general term is $3n - 1$.

9. Consider a well-shuffled deck of 52 cards. What is the probability
    that you would draw a heart or the queen of spades?

10. Six friends are in a flying club but their plane is a two-seater. All
    six show up to fly one sunny Sunday morning. They decide to select
    two at random by putting all names in a hat and drawing two out.
    What is the probability that Sue and Carl get to fly together?
    (Clue: Calculate $(N)_s$, then find the probability.)

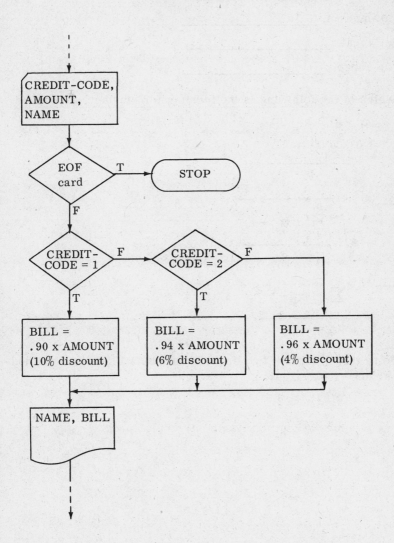

11. For the measurements shown below find:

$$7.5, 9, 14, 10, 7.5, 12, 13, 6, 7.5, 12.5, 12, 8, 11$$

(a) mode _____

(b) median _____

(c) mean _____

(d) range _____

12. Which of the following is the formula for standard deviation?

___(a) $$\dfrac{\sum\limits_{i=1}^{n} |x_i - \bar{x}|}{n}$$

___(b) $$\sqrt{\dfrac{\sum\limits_{i=1}^{n} (x_i - \bar{x})^2}{n}}$$

___(c) $$\dfrac{\sum\limits_{i=1}^{n} (x_i - \bar{x})^2}{n}$$

13. Solve this system of two linear equations. Use either method—comparison or substitution.

$$5x + 6y = 12$$

$$x - 4y = 5$$

14. The value of the determinant of the matrix below is _____.

$$\begin{pmatrix} 4 & 3 \\ -1 & 1 \end{pmatrix}$$

15. A three-by-three matrix multiplied by the three-by-three identity matrix produces:

____(a) the reciprocal of the original matrix
____(b) the identity matrix
____(c) an inverted matrix
____(d) the original matrix

16. Which of the two matrix games illustrated below is strictly determined? _____ Is this a fair game? _____

(a)

4	-4
3	-2

(b)

4	-1
-3	0

ANSWERS TO FINAL TEST
(with solutions and chapter references)

1. (a) 101

$$1 \times 2^0 = 1$$
$$1 \times 2^2 = 4$$
$$1 \times 2^5 = 32$$
$$1 \times 2^6 = \underline{64}$$
$$101$$

(Chapter 1)

(b) $452\frac{1}{8}$

$$1 \times (\tfrac{1}{8})^{-1} = \tfrac{1}{8}$$
$$4 \times 8^0 = 4$$
$$7 \times 8^2 = \underline{448}$$
$$452\tfrac{1}{8}$$

(Chapter 2)

(c) 3053

$$13 \times 16^0 = 13$$
$$14 \times 16^1 = 224$$
$$11 \times 16^2 = \underline{2816}$$
$$3053$$

(Chapter 2)

2. (a) 11001000_2

$$\begin{array}{r} 200 \\ -\ 128 \quad (2^7) \\ \hline 72 \\ -\ 64 \quad (2^6) \\ \hline 8 \\ -\ 8 \quad (2^3) \\ \hline 0 \end{array}$$

(Chapter 1)

(b) $111\ 000\ 100.\ 001_2$

(Chapter 2)

(c) $1011\ 1110\ 1101_2$

(Chapter 2)

3. (a) 1001000_2

$$\begin{array}{r} 1100 \\ \times\ 110 \\ \hline 11000 \\ 1100 \\ \hline 1001000 \end{array}$$

(Chapter 1)

(b) $D719_{16}$

$$\begin{array}{r} B\ 7\overset{+1}{}3\ 9 \\ +\ 1\ F\ E\ 0 \\ \hline D\ 7\ 1\ 9 \end{array}$$

(Chapter 2)

(c) 9759_{16}

$$\begin{array}{r} A\ B\ 7\ 3\ 9 \\ -\ 1\ F\ E\ 0 \\ \hline 9\ 7\ 5\ 9 \end{array}$$

(Chapter 2)

4. (a) F   (~q is false)

   (b) T   (p is true)

   (c) T   (p is true)

   (d) F   (~q is false)

   (e) T $(\sim q \rightarrow p)$

   (f) F $(\sim(\sim q) \rightarrow \sim p,\ \text{or}\ q \rightarrow \sim p)$ (Chapter 3)

5. (a) HARRY FREEMAN   96.00
      (Credit-code of 3 resulted in 4% discount.)

   (b) ANGELA HOOGTERP   45.00
      (Credit-code of 1 resulted in a 10% discount.)   (Chapter 4)

6. (a) $.3479 \times 10^6$
$$.373\ \times 10^6$$
$$+ .0019 \times 10^6$$
$$.3749 \times 10^6$$

   (b) $.469 \times 10^6$, or $.00469 \times 10^8$ (Chapter 5)

7. $87.86

Balance	Period	Interest
700.00	1	21.00
721.00	2	21.63
742.63	3	22.28
764.91	4	22.95
		87.86

                                                    (Chapter 6)

8. $2 + 5 + 8 + 11$

$$3(1) - 1 = 2$$
$$3(2) - 1 = 5$$
$$3(3) - 1 = 8$$
$$3(4) - 1 = 11$$

                                                      (Chapter 7)

9. $7/26$

$$p(\heartsuit) = 13/52$$
$$p(Q\ ) = \ 1/52$$
$$14/52 = 7/26$$

                                                      (Chapter 8)

10. $1/30$ $(N)_s = (6)_2 = 6 \times 5 = 30$ (Chapter 8)

11. (a) 7.5   (occurs 3 times)

   (b) 10

   (c) 10   (130/13)

   (d) 8 $(14 - 6)$ (Chapter 9)

12. (b)   (Chapter 9)

13. $x = 3$, $y = -\frac{1}{2}$ (1) $5x + 6y = 12$

        (2) $x - 4y = 5$

Solve (2) for x:

Substitute in (1): $5(4y + 5) + 6y = 12$

$20y + 25 + 6y = 12$

$26y = -13$

Solve for y: $y = -\frac{1}{2}$

Substitute in (2): $x - 4(-\frac{1}{2}) = 5$

$x + 2 = 5$

Solve for x: $x = 3$ (Chapter 10)

14. 7 $4 - (-3) = 4 + 3$ (Chapter 11)

15. (d)   (Chapter 11)

16. (a)   (−2 is the minimum of a row and maximum of a column.)

        no   (A strictly determined game is fair only if the determining
                value is 0. Here it is −2.)   (Chapter 12)

Index